METHODS AND PROTOCOLS IN FOOD SCIENCE

Series Editor
Anderson S. Sant'Ana
University of Campinas
Campinas, Brazil

For further volumes:
http://www.springer.com/series/16556

Methods and Protocols in Food Science series is devoted to the publication of research protocols and methodologies in all fields of food science.

Volumes and chapters will be organized by field and presented in such way that the readers will be able to reproduce the experiments in a step-by-step style. Each protocol will be characterized by a brief introductory section, followed by a short aims section, in which the precise purpose of the protocol will be clarified.

Basic Protocols in Starch Characterization

Edited by

Maria Teresa Pedrosa Silva Clerici

Department of Food Science and Nutrition, Faculty of Food Engineering, Universidade Estadual de Campinas, Campinas, Brazil

Pedro H. Campelo

Department of Agriculture Engineering and Soils, Federal University of Amazonas, Manaus, Brazil

Humana Press

Editors
Maria Teresa Pedrosa Silva Clerici
Department of Food Science and Nutrition
Faculty of Food Engineering
Universidade Estadual de Campinas
Campinas, Brazil

Pedro H. Campelo
Department of Agriculture Engineering and Soils
Federal University of Amazonas
Manaus, Brazil

ISSN 2662-950X ISSN 2662-9518 (electronic)
Methods and Protocols in Food Science
ISBN 978-1-0716-5115-5 ISBN 978-1-0716-5116-2 (eBook)
https://doi.org/10.1007/978-1-0716-5116-2

This Humana imprint is published by the registered company Springer Science+Business Media, LLC, part of Springer Nature.
The registered company address is: 1 New York Plaza, New York, NY 10004, U.S.A.

Preface to the Series

Methods and Protocols in Food Science series is devoted to the publication of research protocols and methodologies in all fields of food science. The series is unique as it includes protocols developed, validated and used by food and related scientists as well as theoretical basis are provided for each protocol. Aspects related to improvements in the protocols, adaptations and further developments in the protocols may also be approached.

Methods and Protocols in Food Science series aims to bring the most recent developments in research protocols in the field as well as very well established methods. As such the series targets undergraduate, graduate and researchers in the field of food science and correlated areas. The protocols documented in the series will be highly useful for scientific inquiries in the field of food sciences, presented in such way that the readers will be able to reproduce the experiments in a step-by-step style.

Each protocol will be characterized by a brief introductory section, followed by a short aims section, in which the precise purpose of the protocol is clarified. Then, an in-depth list of materials and reagents required for employing the protocol is presented, followed by a comprehensive and step-by-step procedures on how to perform that experiment. The next section brings the do's and don'ts when carrying out the protocol, followed by the main pitfalls faced and how to troubleshoot them. Finally, template results will be presented and their meaning/conclusions addressed.

The *Methods and Protocols in Food Science* series will fill an important gap, addressing a common complain of food scientists, regarding the difficulties in repeating experiments detailed in scientific papers. With this, the series has a potential to become a reference material in food science laboratories of research centers and Universities throughout the world.

Campinas, Brazil ***Anderson S. Sant'Ana***

Preface

Starch is one of the most widely studied and applied biopolymers in the food, pharmaceutical, and industrial sectors. Its structural complexity, functional versatility, and economic relevance make it a constant subject of innovation and methodological refinement.

This book presents a comprehensive and updated collection of analytical methods for starch characterization, with contributions from experienced researchers in the field. The techniques covered include classical and emerging approaches for evaluating physicochemical properties, structural features, and functional behavior.

The methodologies compiled herein were developed and refined by authors from diverse countries and scientific traditions, each contributing valuable perspectives and practical insights. Such global collaboration enriches the technical depth of this work and ensures that the approaches described are relevant, adaptable, and reflective of the current state of starch analysis worldwide.

Adherence to standardized protocols in starch analysis is essential for scientific accuracy and the rational use of resources. Well-established procedures allow researchers to optimize their time, minimize waste of chemical and enzymatic reagents, and promote more efficient use of laboratory glassware and equipment. These practices contribute to more sustainable research environments and reduce experimental variability.

Moreover, the use of consistent methodologies enables the generation of data that can be reliably compared across studies. This comparability strengthens scientific communication, facilitates meta-analyses, and supports the construction of robust knowledge bases in starch science. Reliable methods are the foundation for reproducibility and progress in research and development.

The study of new sources of starch, such as seeds from fruit processing facilities, non-industrialized roots and tubers, and young bamboo culm, is a somewhat arduous process due to lack of previously published extraction methods. Therefore, in addition to the basic analysis of the chemical composition of starch (Chaps. 1–5), we also present the analysis of total starch content using infrared methodology, which requires only a minimum amount of sample (Chap. 9). Morphological analysis methods are presented in Chaps. 14 and 15, with the majority being performed by physical tests, following the current analytical trend, in order to not generate effluents. As the number of studies on new types of starch increases, some authors seek to know whether the studied starch is resistant or if it produces prebiotic effects; therefore, Chap. 18 presents the methodology to do so.

Native starches can be extracted in a fully watery medium (Chap. 6) as a way to avoid modifying it. However, if the sample presents high protein content, enzymatic darkening effect enzymes, and sugars that induce fermentation during the decantation processes at the laboratory, the method that uses metabisulfite or its derivative salts is presented as an alternative (Chap. 7). Chapters 10 and 11 present protocols to determine amylose and amylopectin, which is very important to know if the starch is native or modified (Chap. 11) and to predict the behavior of the starch in products, as paste (Chap. 12); syneresis and retrogradation (Chap. 13); and rheology (Chaps. 16 and 17). Protocols of techno-functional properties such as water absorption and water solubility, oil binding, and emulsifying properties (Chaps. 20–22) were used only for proteins; however, with a lot of new types of starch and many modified starches, these protocols have become important to know the application of starch in various products. Lastly, two emergent starch modification methods are presented in Chaps. 23 (starch nanofibers) and 24 (starch nanocapsules), produced by electrospinning process.

On a personal note, the journey to completing this book has been one of significant challenges. From health struggles to professional setbacks, there were many moments when the path seemed insurmountable. However, this work would not have been completed without the unwavering support of our colleagues, the invaluable contributions from all the authors, and the deep understanding and encouragement from Mrs. Monica Suchy, our fellow editor. Their dedication and belief in this project helped us move forward when we thought we could not. For this, we are deeply grateful.

Our aim with this volume is to provide students, researchers, and industry professionals with a practical and reliable reference, not only for academic research but also for quality control and product development. The diversity of methods presented reflects the multidisciplinary nature of starch science and the importance of methodological rigor in advancing the understanding of this essential biomaterial.

As editors of this book, we are deeply grateful to all the contributors who generously shared their expertise, and to the reviewers and collaborators who supported the development of this work. We hope this book will serve as a valuable tool and inspire new investigations in starch analysis.

Campinas, Brazil *Dr. Maria Teresa Pedrosa Silva Clerici*

Contents

Contributors

Olaide A. Akintayo • *Department of Home Economics and Food Science, University of Ilorin, Ilorin, Nigeria; Discipline of Food Science, School of Agriculture, Food and Wine, The University of Adelaide, Glen Osmond, Adelaide, SA, Australia*

Kivia Mislaine Albano • *Center for Arts, Humanities and Technology (CAHT), Federal University of São Carlos (UFSCar), São José do Rio Preto, Brazil*

Nuria Aleixos-Borrás • *Universidad Politecnica de Valencia (UPV), Valencia, Spain*

Pedro Esteves Duarte Augusto • *Université Paris-Saclay, CentraleSupélec, Laboratoire de Génie des Procédés et Matériaux, Centre Européen de Biotechnologie et de Bioéconomie (CEBB), Pomacle, France*

Amanda Teixeira Badaró • *Department of Food Engineering, Universidade Estadual de Campinas, Campinas, Brazil*

Douglas Fernandes Barbin • *Department of Food Engineering, Universidade Estadual de Campinas, Campinas, Brazil*

Gabriela Noel Barrera • *Facultad de Ciencias Exactas, Físicas y Naturales - Universidad Nacional de Córdoba, Córdoba, Argentina*

Jaqueline de Araújo Bezerra • *Coordination of Technology and Innovation (COTEI), National Institute for Amazonian Research (INPA), Manaus, Amazonas, Brazil*

Jose Blasco • *Instituto Valenciano de Investigaciones Agrarias (IVIA), Valencia, Spain*

Natali Alcantara Brandão • *Department of Food Science and Nutrition, Faculty of Food Engineering, Universidade Estadual de Campinas, Campinas, Brazil; Department of Food and Nutrition, Federal University of Santa Maria, Palmeira das Missões Campus, Palmeira das Missões, RS, Brazil*

Pedro Henrique Campelo • *Department of Food Technology, Federal University of Viçosa (UFV), Viçosa, Minas Gerais, Brazil*

Hugo José Martins Carvalho • *Department of Food Science and Nutrition, Faculty of Food Engineering, Universidade Estadual de Campinas, Campinas, Brazil*

Cinthia Baú Betim Cazarin • *Food Science and Nutrition Department, Faculty of Food Engineering, Universidade Estadual de Campinas, Campinas, Brazil*

Maria Teresa Pedrosa Silva Clerici • *Department of Food Science and Nutrition, Faculty of Food Engineering, Universidade Estadual de Campinas, Campinas, Brazil; Department of Food and Nutrition, Federal University of Santa Maria, Palmeira das Missões Campus, Palmeira das Missões, RS, Brazil*

Josias Martins dos Anjos Cruz • *Graduate Program in Chemistry, Federal University of Amazonas (UFAM), Manaus, Amazonas, Brazil*

Luis Jam Pier Cruz-Tirado • *Department of Food Engineering, Universidade Estadual de Campinas, Campinas, Brazil*

Estefania Julia Dierings de Souza • *Department of Agroindustrial Science and Technology, Laboratory of Biopolymers and Nanotechnology in Food, Federal University of Pelotas, Pelotas, Brazil*

Olayemi E. Dudu • *Department of Food Science and Technology, Bells University of Technology, Ota, Nigeria*

MÁRIA HERMINIA FERRARI FELISBERTO • *Department of Food Technology, Federal University of Viçosa, Viçosa, MG, Brazil*
BEATRIZ MANFRINATO FERREIRA • *Department of Food Science and Nutrition, Faculty of Food Engineering, Universidade Estadual de Campinas, Campinas, Brazil*
SUNGIL FERREIRA • *University of Arkansas System Division of Agriculture - Department of Food Science, Fayetteville, AR, USA*
LAURA MARTINS FONSECA • *Department of Agroindustrial Science and Technology, Laboratory of Biopolymers and Nanotechnology in Food, Federal University of Pelotas, Pelotas, Brazil*
MARINA VILAR GERALDI • *Department of Food Science and Nutrition, Faculty of Food Engineering, Universidade Estadual de Campinas, Campinas, Brazil*
ALVARO RENATO GUERRA DIAS • *Department of Agroindustrial Science and Technology, Laboratory of Biopolymers and Nanotechnology in Food, Federal University of Pelotas, Pelotas, Brazil*
DÂMARIS CARVALHO LIMA • *Department of Agri-food Industry, Food and Nutrition (LAN), Luiz de Queiroz College of Agriculture (ESALQ), University of São Paulo (USP), Piracicaba, Brazil*
BIANCA CHIEREGATO MANIGLIA • *São Carlos Institute of Chemistry (IQSC), University of São Paulo (USP), Sao Carlos, Brazil*
MÁRIO ROBERTO MARÓSTICA JUNIOR • *Department of Food Science and Nutrition, Faculty of Food Engineering, Universidade Estadual de Campinas, Campinas, Brazil*
POLIANA MOSER • *Department of Food Engineering and Technology, Institute of Biosciences, Humanities and Exact Sciences, São Paulo State University, São José do Rio Preto, SP, Brazil*
ELISA CRISTINA ANDRADE NEVES • *School of Food Engineering, Federal University of Pará, Belém, Brazil*
ROBERTO DE FREITAS NEVES • *School of Chemical Engineering, Federal University of Pará, Belém, Brazil*
VÂNIA REGINA NICOLETTI • *Department of Food Engineering and Technology, Institute of Biosciences, Humanities and Exact Sciences, São Paulo State University, São José do Rio Preto, SP, Brazil*
SAMSON A. OYEYINKA • *Department of Home Economics and Food Science, University of Ilorin, Ilorin, Nigeria; Centre for Innovative Food Research (CIFR), Department of Biotechnology and Food Technology, University of Johannesburg, Doornfontein Campus, Johannesburg, South Africa; School of Agri-Food Technology and Manufacturing, College of Health and Science, University of Lincoln, Lincoln, UK*
ELDER PACHECO DA CRUZ • *Department of Agroindustrial Science and Technology, Laboratory of Biopolymers and Nanotechnology in Food, Federal University of Pelotas, Pelotas, Brazil*
IBRAHIM PALABIYIK • *Department of Food Engineering, Faculty of Agriculture, Tekirdag Namik Kemal University, Tekirdag, Türkiye*
PABLO MARTIN PALAVECINO • *Facultad de Ciencias Exactas, Físicas y Naturales, Universidad Nacional de Córdoba, Córdoba, Argentina*
ELIENE PENHA RODRIGUES PEREIRA • *Department of Food Science and Nutrition, Faculty of Food Engineering, Universidade Estadual de Campinas, Campinas, Brazil*
NASIM KIAN POUR • *Gastronomy and Culinary Arts Department, Küçükçekmece, Istanbul Aydın University, Istanbul, Türkiye*

Pablo Daniel Ribotta • *Facultad de Ciencias Exactas, Físicas y Naturales, Universidad Nacional de Córdoba, Córdoba, Argentina; Instituto Superior de Investigación, Desarrollo y Servicios en Alimentos, Secretaria de Ciencia y Tecnologia, Universidad Nacional de Córdoba, Juan Filloy s/n - Ciudad Universitaria, Córdoba, Argentina*

Meliza Lindsay Rojas • *Dirección de Investigación, Innovación y Responsabilidad Social, Universidad Privada del Norte, Trujillo, Perú*

Rebeca Salvador-Reyes • *Facultad de Ingeniería, Universidad Tecnológica del Perú, Lima, Perú*

Edgar Aparecido Sanches • *Laboratory of Nanostructured Polymers (NANOPOL), Federal University of Amazonas, Manaus, Amazonas, Brazil*

Anderson S. Sant'Ana • *Department of Food Science and Nutrition, Faculty of Food Engineering, Universidade Estadual de Campinas, Campinas, Brazil*

Bruna Lago Tagliapietra • *Department of Food and Nutrition, Federal University of Santa Maria, Palmeira das Missões Campus, Palmeira das Missões, RS, Brazil; Department of Food Science and Nutrition, Faculty of Food Engineering, Universidade Estadual de Campinas, Campinas, SP, Brazil*

Omer Said Toker • *Department of Food Engineering, Faculty of Chemical and Metallurgical Engineering, Yildiz Technical University, Istanbul, Türkiye*

Elessandra da Rosa Zavareze • *Department of Agroindustrial Science and Technology, Laboratory of Biopolymers and Nanotechnology in Food, Federal University of Pelotas, Pelotas, Brazil*

Chapter 1

Moisture

Bruna Lago Tagliapietra, Natali Alcantara Brandão and Maria Teresa Pedrosa Silva Clerici

Abstract

The chapter addresses the importance of determining moisture content in starches extracted from various food sources such as grains, tubers, roots, and others. The common method for moisture determination is drying in an oven at 105°C until the mass loss stabilizes. However, different methods for moisture determination are mentioned, including oven drying, infrared radiation drying, and Karl Fischer titration and can be used according to laboratory availability.

Keywords Starch, Stability, Oven, Drying, Water activity

1 Introduction

Moisture content refers to the mass loss that occurs when starch is subjected to water removal processes, which can be through heat, solvent use, and lyophilization. When starch is extracted and purified from its food matrix, it can be dehydrated to reach an equilibrium moisture content, allowing safe storage for extended periods. The equilibrium moisture content for cereal starches ranges from 10% to 13%, whereas for tuber and root starches, it is 14% to 18%.

The most common method is heating the sample in an oven with or without air circulation at 105°C until the mass loss becomes constant. Some authors have used the same method as AACCI [1] for flours, which is heating in an oven at 130°C for 1 h. However, for starches extracted in experimental and nonoptimized processes, excessive browning reactions and mass loss may occur, compromising the result. For the study of native starches, where the effect of heat treatment is not desired as a modifying agent, it is necessary to resort to vacuum ovens, where the pressure is reduced, and the temperature is maintained around 70°C or less, when the starch gelatinization temperature is known. This reduced temperature procedure preserves

Maria Teresa Pedrosa Silva Clerici and Pedro H. Campelo (eds.), *Basic Protocols in Starch Characterization*, Methods and Protocols in Food Science, https://doi.org/10.1007/978-1-0716-5116-2_1,

sample integrity and prevents surface reactions that could hinder water evaporation.

The main methods for moisture determination in starchy matrices are:

(a) Use of drying ovens → This method is the most commonly used for starches and is based on water removal through direct heating, with hot air absorbed by a thin layer of starch and then conducted into the sample's interior. This method typically lasts for about 6–18 h, requiring several weighings of the sample until it reaches a constant weight.

(b) Infrared radiation drying → This method is faster and involves heat penetration into the sample. The method utilizes an infrared radiation lamp with 250–500 watts, with the filament reaching a temperature between 2000 and 2500°K (700°C) [2]. The drying time for starches should be predetermined, using a temperature of 105°C and creating a mass loss curve for 15–20 min until the mass remains constant for 20–30 s; this time is noted and should be used for subsequent analyses of the same type of starch. Excessive heating times lead to sample darkening or burning, compromising results. For instance, cereal and tuber starches may require different drying times for moisture determination.

(c) Karl Fischer titration → In this methodology, the sample is diluted, usually in methanol, with the Karl Fischer reagent, which is a solution containing amine, iodine, and sulfur dioxide. The measurement is carried out by the equipment employing platinum electrodes and utilizing calcium carbide ($CaCl_2$) or calcium hydride (CaH_2) [2]. This method is used in the pharmaceutical and cosmetic industries, where starch acts as an excipient for active substances, measuring moisture variations from 0.01% to 100%, with high precision and accuracy, also applicable in nonaqueous media. It is a rapid method used to monitor dehydration processes.

It is important to follow appropriate standards and procedures as different sample types may require method adjustments, such as specific drying times. This protocol describes the general method for moisture determination in a chemical analysis laboratory.

2 Material

1. Porcelain crucibles (15–25 mL capacity) or glass, aluminum, or stainless-steel Petri dishes sized 90 × 15 mm. Strictly follow the desiccator usage procedure (*see* **Notes 1–3**);
2. Spatula with a 12–18-cm stainless-steel blade and spoon;
3. Starch sample of 50–100 g, identified by code, which can consist of 3 letters or digits;

4. Tongs for crucibles, size 22, zinc-plated iron, for placing and removing crucibles in the oven, desiccator, and analytical balance;
5. Long-cuffed thermal gloves with nonslip palms for removing samples from the oven using tongs;
6. Permanent marker pen;
7. A Spreadsheet for recording weighings can be a laboratory notebook or a computer;
8. Common PPE for a chemical analysis laboratory.

3 Method

Determination of starch moisture content by heating at 105°C in an oven until constant weight [1–6].

3.1 Equipment

1. Drying and sterilization oven without or with air circulation reaching at least 130°C or a vacuum oven;
2. Desiccator containing predried silica gel, with a lid, 30 cm in diameter or larger, according to the number of samples to be analyzed, as shown in Fig. 1A, B;
3. Analytical balance with a precision of up to 0.0001 g (1 mg);

3.2 Procedures

3.2.1 Preanalysis Procedures

1. Preseparate the required number of containers (2.1, item 1) according to the number of samples and perform the analyses in triplicate or more. For example, if two samples are being analyzed, it is necessary to separate six containers. Check if

Fig. 1 (**A**) Complete glass desiccator and (**B**) blue silica gel, previously dried in an oven at 105°C. Source: Authors' collection, 2025

they have preexisting identification that withstands washing and heating; if not, use a high-temperature resistant pen for identification;

2. Check the operational procedure of the circulating air oven and ensure it is precalibrated. Note these data. Strictly adhere to the usage procedure;
3. Check the standardized operational procedure of the analytical balance and ensure it is calibrated, noting the calibration date. Strictly adhere to the usage procedure;
4. For data collection, prepare a results table in the laboratory notebook, as shown in Table 1 (*see* **Notes**).

3.2.2 Moisture Determination

1. Wash the containers (2.2, item 1) with appropriate neutral detergent, removing any residues, and then rinse three times with distilled water;
2. Dry the containers (2.2, item 1) in an oven at 130°C for 1 h, remove them from the oven, and place them in the desiccator (*see* **Note 4**);
3. Cool the crucibles in the desiccator for 40 min, not exceeding 1 h, to proceed to step 4;
4. Using tongs, weigh the crucible on the analytical balance; in Table 1, record the weight of the empty crucible (P1) in the corresponding replica number;
5. Weigh approximately 2.0000 g of starch sample with the precision of the analytical balance; in Table 1, record the weight of the wet sample (P2) (*see* **Notes 5–7**);
6. Place the crucibles with the starch samples in the oven at 105°C and wait for 6–7 h for the first weighing.
7. Remove the crucibles from the oven and place them in the desiccator to cool to room temperature (25°C) for 40 min, not exceeding 60 min;
8. Perform the first weighing, record the mass (crucible + starch sample) (P3);
9. Return the crucible with the sample to the oven for another 2 h;
10. Repeat steps 12 and 13, conducting the second weighing (P4).
11. Return the crucible to the oven (step 14) and repeat steps 12 and 13 to obtain the third weighing (P5) and, if necessary, the fourth weighing (P6);
12. Evaluate if the results obtained in P4 and P5 remained constant, with a variation of maximum 0.0010 between them;

Table 1
Spreadsheet for recording collected data and calculating moisture analysis results

Moisture determination (%)													
Sample													
Analyst													
Location and date													
Sample identification (code and reference)	Number	Replicate	Weight of empty crucible (g) (P1)	Initial weight of wet starch (g) (P2)	First weighing weight of crucible + dry starch (g) (P3)	Second weighing weight of crucible + dry starch (g) (P4)	Third weighing weight of crucible + dry starch (g) (P5)	Fourth weighing weight of crucible + dry starch (g) (P6)	Dry matter weight (dry starch) (g) P6 - P1	Moisture (%)	Average (%)	SD	CV
CS*	1	CS1											
	2	CS2											
	3	CS3											

*Example for recording data of a starch sample; *CS*, corn starch; *SD*, standard deviation; *CV*, coefficient of variation

Table 2
Example of results for determining moisture in starches

Moisture determination (%)													
Sample													
Analyst													
Location and date													
Sample identification (code and reference)	Number	Replicate	Weight of empty crucible (g) (P1)	Initial weight of wet starch (g) (P2)	First weighing weight of crucible + dry starch (g) (P3)	Second weighing weight of crucible + dry starch (g) (P4)	Third weighing weight of crucible + dry starch (g) (P5)	Fourth weighing weight of crucible + dry starch (g) (P6)	Dry matter weight (dry starch) (g) P6 - P1	Moisture (%)	Average (%)	SD	CV
	1	CS1	45.3292	2.0362	47.1808	47.1612	47.153	47.1528	1.8236	10.4410			
CS*	2	CS2	40.6456	2.0125	42.4615	42.4408	42.4371	42.4369	1.7913	10.9913	0.1331	0.0092	6.8968
	3	CS3	44.896	2.0321	45.8406	45.8221	45.7157	46.7156	1.8196	10.4572			

*Example for recording data of a starch sample. *CS*, corn starch; *SD*, standard deviation; *CV*, coefficient of variation

13. If the results of the second and third weighing are still not constant, repeat steps 14, 12, and 13 in this sequence, recording the weighing until the weight is constant for at least two consecutive results;
14. Record all values in the data recording spreadsheet (Table 1).

3.3 Results

Calculate the moisture content of the sample according to the following equation:

$$\text{Moisture } (\% \text{ or g/100 g}) = \frac{\text{Weight of wet starch (P2)} - \text{Weight of dry starch}}{\text{Weight of wet starch (P2)}} \times 100$$

Where

- Weight of wet starch (P2): Initial weight of wet starch (g)
- Weight of dry starch: Consists of the remaining dry matter after drying, calculated from the last crucible weighing (after constant weight) as described below
- Weight of dry starch = P5 (weight of crucible + dry starch) – weight of empty crucible

After obtaining the moisture content of the three replicates per sample, calculate the average, standard deviation (SD), and coefficient of variation (CV), as provided by data tabulation programs such as Microsoft Excel or Google Sheets. The standard deviation is considered optimal when its value is as close to zero as possible. A coefficient of variation below 10% indicates a dispersion close to ideal for this type of analysis.

3.4 Presentation of Results

Results should always be presented as the average of the triplicate for each sample, in percentage (%), g/100 g, g/kg, considering four decimal places, plus the calculated standard deviation, as shown in the example below (Table 2). (see **Note 8**, Table 3).

4 Notes

1. Selection of the porcelain crucible will be indicated if the sample proceeds to ash determination.
2. Due to its porous surface, the porcelain crucible is not recommended for fluid products.
3. The aluminum dish cannot be used with acidic pH products such as fermented or modified starches modified starches.

Table 3
Moisture content of corn and potato starch

Sample	Moisture (%)
Corn starch	11.0032 ± 0.0040
Potato starch	12.0025 ± 0.0059

4. Throughout the analysis, always handle the crucibles with the aid of tongs and use thermal gloves to place and remove crucibles from the oven, avoiding transferring moisture and oils from the hands to the container and interfering with the analysis results.
5. The amount of sample used for moisture determination may vary depending on the food matrix being evaluated. For starchy flours, if available, 5–10 g of sample can be used on aluminum plates. However, if it is a work with a scarce matrix, 0.5–2 g of the sample can be used in porcelain crucibles.
6. Record the weights with a precision of at least four decimal places, and the final result can be rounded to two decimal places.
7. When weighing the crucible, record its weight, and then record the weight of the starch, having the weight of the crucible + weight of the starch; or, as shown in Table 1, record the weight of the crucible, tare the balance to zero it, and weigh the starch, recording the weight of the starch.
8. In Table 3, we provide an example of an analysis for obtaining a result with constant weight.

Acknowledgments

The authors would like to thank the National Council for Scientific and Technological Development (CNPq), Brazil, for the scholarships awarded to MTP Clerici (312660/2023-5) and the São Paulo Research Foundation (FAPESP), Brazil (Process Number 2023/12726-8) for the postdoctoral fellowship to BL Tagliapietra. Additionally, the authors express gratitude to the Coordination for the Improvement of Higher Education Personnel (CAPES, financial code 001), Brazil, for the scholarships awarded to NA Brandão (Process 88887.605710/2021-00), and for supporting the Graduate Program in Food Technology at the Universidade Estadual de Campinas (UNICAMP), São Paulo, Brazil.

References

1. American Association of Cereal Chemists (2010) AACCI approved methods of analysis, 11th edn. AACC International, St. Paul
2. Cecchi HM (2003) Theoretical and practical foundations in food analysis. Unicamp, Campinas, p 13
3. Association of Official Analytical Chemists (2006) AOAC, 18th edn. Association of Official Analytical Chemists, Washington, DC
4. Instituto Adolfo Lutz (2008) IAL—Anal Stand Adolfo Lutz Inst Physicochemical methods food Anal, 4th edn. São Paulo, São Paulo, p. 1020
5. Oliveira LA (2010) Laboratory manual: physicochemical analyses of fruits and cassavas, 1st edn. Embrapa Cassava and Fruit, Cruz das Almas, p 248
6. Silva CO, Tassi EMM, Pascoal GB (2016) Food science: principles of bromatology, 1st edn. Rubio, Rio de Janeiro

Check for updates

Chapter 2

Ash

Natali Alcantara Brandão, Bruna Lago Tagliapietra and Maria Teresa Pedrosa Silva Clerici

Abstract

Ash in foods, also referred to as fixed mineral residue, represents the inorganic residues remaining after the complete oxidation of organic matter through combustion, obtained by incineration at about 550–570°C for 4 h or more, provided that the resulting ash is white or grayish in color and without black spots. The presence of mineral salts in starches may indicate the degree of purity, with quantities below 0.1%–0.2%.

Keywords Starch, Mineral residue, Inorganic residue, Muffle furnace

1 Introduction

The ash content of a food corresponds to the inorganic residues remaining after the combustion of organic matter, which is converted into CO_2, H_2O, and NO_2. The obtaining of ash, or residue by incineration, involves heating the starch sample in a muffle furnace at a temperature of approximately 550–570°C for a period exceeding 4 h or until complete combustion of the organic matter occurs, i.e., until a constant mass is obtained. Therefore, the process must be carried out in stages, with regular weighings after the burning period. This process results in the formation of ashes with a uniform color, generally white or grayish [3].

Ash is mainly composed of potassium, sodium, calcium, and magnesium salts, with small amounts of other minerals. Thus, its determination is an indication of starch purity and serves as a starting point for the analysis of specific minerals. For example, in phosphated starches, whether natural, like potato starch, or chemically modified, the study of phosphorus and its binding to starch can be conducted through thermogravimetric analysis and P31-NMR.

Maria Teresa Pedrosa Silva Clerici and Pedro H. Campelo (eds.), *Basic Protocols in Starch Characterization*, Methods and Protocols in Food Science, https://doi.org/10.1007/978-1-0716-5116-2_2,

2 Material

1. Porcelain crucibles.
2. Spatula.
3. Starch sample.
4. Long-handled tongs for use in muffle furnaces.
5. Thermal gloves.
6. Standard laboratory PPE for chemical analysis.

3 Method

Determination of ash by muffle furnace incineration [1–3].

3.1 Equipment

1. Muffle furnace reaching 550°C.
2. Desiccator containing predried silica gel.
3. Analytical balance with precision up to 0.0001 g (1 mg).
4. Fume hood.
5. Sand bath or heating plate.

3.2 Procedures

1. Identify the bottom of the crucibles using a heat-resistant graphite pencil.
2. Wash the crucibles with appropriate neutral detergent, removing any residue, then rinse three times with distilled water.
3. Calcine the porcelain crucible, i.e., subject it to burning in the muffle furnace at 550°C for 2–3 h, cool down to 300°C; once the thermostat indicates 300°C, turn off the muffle furnace and wait for the temperature to drop to 150°C, remove the porcelain crucible from the furnace when it reaches a temperature of 100°C, and place it in a desiccator containing silica gel for 40 min to 1 h maximum until cooled.
4. Weigh and record the weight of the empty crucible (P1).
5. Weigh approximately 2.0000 g of the sample and record the weight (P2).
6. Initiate the initial combustion of the material using a sand bath or heating plate; this procedure must be carried out within the fume hood, and the combustion should be slow, between 100°C and 130°C, until a charcoal-like mass forms in the sample.
7. Place the crucibles (after step 6) with the samples in the muffle furnace initially at 550°C for 3–4 h.

Table 1
Spreadsheet of data records collected in the analysis

Ash determination (%)										
Sample										
Analyst										
Date										
Sample identification (code and reference)	Number	Replicate	Weight of empty crucible (g) (P1)	Sample weight (g) (P2)	Weight of crucible + ash (g) (P3)	Weight of ash (g)	Ash content (%)	Average (%)	SD	CV
	1	CS1								
CS*	2	CS2								
	3	CS3								

*Example for recording data of a starch sample. *CS*, corn starch; *SD*, standard deviation; *CV*, coefficient of variation

Table 2
Example of results for determining ash in starches

Ash determination (%)										
Sample										
Analyst										
Date										
Sample identification (code and reference)	**Number**	**Replicate**	**Weight of empty crucible (g) (P1)**	**Sample weight (g) (P2)**	**Weight of crucible + ash (g) (P3)**	**Weight of ash (g)**	**Ash content (%)**	**Average (%)**	**SD**	**CV**
CS*	1	CS1	45.3292	2.0362	45.3318	0.0026	0.1277	0.1331	0.0092	6.8968
	2	CS2	40.6456	2.0125	40.6481	0.0025	0.1242			
	3	CS3	44.8960	2.0321	44.8985	0.0025	0.1230			

*Example for recording data of a starch sample. *CS*, corn starch; *SD*, standard deviation; *CV*, coefficient of variation

8. Upon reaching the set time, program the muffle furnace to cool down to 300°C; once the thermostat indicates 300°C, turn off the muffle furnace and wait for the temperature to drop to 150°C; using long-handled thermal gloves and tongs, remove the crucibles and place them in the desiccator, previously prepared.
9. Check if the obtained ash is white at the end of the process; if yes, proceed to step 10, if not, proceed with a new combustion for an additional 3–4 h, or you can proceed to step 10 if you want to expedite the process using a chemical reagent.

   - If the ash has black spots, add two to three drops of concentrated nitric acid (HNO_3) or 30% (m/v) hydrogen peroxide (H_2O_2), and return to the muffle furnace at 550°C for one more hour. Starches containing phosphate groups generally exhibit black spots in the ash.

10. Weigh and record the mass (crucible + sample) (P3).
11. Record all values in the data recording spreadsheet (see Tables 1 and 2).

4 Results

Calculate the ash content of the sample according to the following equation

$$\text{Ash}\left(\%\text{ or }g/100\text{g}\right)=\frac{\text{Weight of ash}}{\text{Weight of sample (P2)}}\times 100$$

Where

- Weight of ash (g) = P3 (Weight of crucible + ash) – P1 (Weight of empty crucible);
- P2: Initial weight of the sample (g).

Acknowledgments

The authors would like to thank the National Council for Scientific and Technological Development (CNPq), Brazil, for the scholarships awarded to MTP Clerici (312660/2023-5) and the São Paulo Research Foundation (FAPESP), Brazil (Process Number 2023/12726-8) for the postdoctoral fellowship to BL Tagliapietra. Additionally, the authors express gratitude to the Coordination for the Improvement of Higher Education Personnel (CAPES, financial code 001), Brazil, for the scholarships awarded to NA Brandão (Process 88887.605710/2021-00), and for supporting the Graduate Program in Food Technology at the Universidade Estadual de Campinas (UNICAMP), São Paulo, Brazil.

References

1. American Association of Cereal Chemists (2010) AACCI approved methods of analysis, 11th edn. AACC International, St. Paul
2. Association of Official Analytical Chemists (2006) AOAC, 18th edn. Association of Official Analytical Chemists, Washington, DC
3. Instituto Adolfo Lutz (2008) IAL—analytical standards of the Adolfo Lutz Institute. Physicochemical methods for food analysis, 4th edn. São Paulo, São Paulo, p 1020

Chapter 3

Lipids

Bruna Lago Tagliapietra, Natali Alcantara Brandão and Maria Teresa Pedrosa Silva Clerici

Abstract

Lipids are a group of chemical compounds characterized by their insolubility in water and solubility in organic solvents. Lipid analysis is significant in the food industry to assess the purity of starch and flours since lipids are susceptible to chemical modifications during processing, affecting sensory, nutritional, and functional properties. The main methods for determining lipids in starchy matrices include hot extraction by Soxhlet, using petroleum ether or ethyl ether, and cold extraction, known as the Bligh & Dyer method, involving chloroform–methanol–water. These methods provide crucial information to ensure the quality and purity of food products.

Keywords Extraction, Starch, Bligh & Dyer, Lipids

1 Introduction

Lipids are components insoluble in water and soluble in organic solvents, such as ethyl ether, petroleum ether, acetone, chloroform, benzene, and alcohols. These nonpolar solvents extract the neutral lipid fraction, including free fatty acids, mono-, di-, and triglycerides, and some more polar compounds like phospholipids and glycolipids [1]. Routine determination of lipids involves extracting the ether extract using organic solvents, providing hot or cold extraction of the lipid fraction from the sample.

Determining the lipid content in starch is essential to assess the purity of flours and serves as a fundamental parameter for the food industry. Lipids are subject to various chemical alterations during food processing, storage, and consumption, generating undesirable substances and altering several properties such as sensory quality, nutritional value, and functionality [2]. In starches, lipids may be present complexed with amylose if the raw material (grains, seeds, and among others) contains lipids and phospholipids and undergoes maceration in water, as in the extraction of corn starch, where residual lipid

Maria Teresa Pedrosa Silva Clerici and Pedro H. Campelo (eds.), *Basic Protocols in Starch Characterization*, Methods and Protocols in Food Science, https://doi.org/10.1007/978-1-0716-5116-2_3,

remains in the corn, approximately 0.3%. However, there are modified starchesmodified starches using organic compounds, such as OSA-starch, where the ester bond must be broken to determine the lipid content, which may be bound but also may have a free percentage.

The main methods for determining lipids in starchy matrices are:

(a) Hot solvent lipid extraction → This method relies on the exhaustive extraction of lipids from the sample using an organic solvent in a Soxhlet apparatus and is performed in three stages. Initially, fat is extracted from the sample using solvents, followed by solvent removal through evaporation and drying to quantify the fat content of the sample. The solvent used will depend on the lipid components present in the sample, with the two most commonly used for hot extractions being petroleum ether, ethyl ether [1], and hexane. It is important to note that in this case, it is usually stated that the determined value is in ether extract or hexane extract. When using ether, it is important to know that petroleum ether is a mixture of hydrocarbons, and ethyl ether is a compound of the ether group, so extractions may result in different contents and compounds depending on the solvent used, which should be highlighted in the methodology description. Ideally, the sample should be dried before analysis, as this allows the solvent to penetrate more easily into the sample. The efficiency of the extraction depends on several factors, including the nature of the sample, particle size (smaller particles facilitate solvent penetration), moisture (sample water hinders solvent penetration), the existence of lipid binding with other sample compounds, and the relative amount of solvent-sample [2].

(b) Cold solvent lipid extraction → This method, known as the Bligh & Dyer [3], is based on lipid extraction by mixing three solvents: chloroform–methanol–water, and has the advantage of extracting all classes of lipids, both polar and nonpolar. The sample is initially mixed with methanol and chloroform in a proportion that forms a single phase with the sample, then more chloroform and water are added to form two phases: one of chloroform containing lipids and another of methanol plus water, containing nonlipid substances. Subsequently, the chloroform phase with lipids is isolated, and after evaporation, the amount of fat in the sample is obtained [1, 3]. Cold extraction has the advantage of not requiring specialized and sophisticated equipment.

2 Material

- Falcon tube of 50 mL with cap.
- Falcon tube of 15 mL with cap.
- Spatula.

- Beaker of 20 mL.
- Beaker de 10 mL.
- Filter papers of 90 mm.
- Funnel.
- volumetric pipette of 5 mL.
- Graduated pipettes of 10 and 20 mL.
- Dispensers for pipettes.
- Falcon pipe shelf.

2.1 Reagents

- Methanol (P.A.).
- Chloroform (P.A.).
- Anhydrous Sodium Sulfate (P.A.).
- 1.5% Sodium sulfate solution.

3 Method

Lipid determination Bligh & Dyer [3]: The principle of this method utilizes an initial ratio, which must be strictly followed, of 1:2:0.8 of chloroform:methanol:water, where the three solvents coexist in a homogeneous mixture. The addition of more chloroform and more water changes the ratio to 2:2:1.8, causing the complete separation of chloroform carrying the lipids (bottom layer), thus all lipids from the sample are dissolved in 20 mL of chloroform [4, 5].

3.1 Equipment

1. Analytical balance.
2. Mechanical tube shaker (20 rpm).
3. Rotary centrifuge.
4. Fume hood.
5. Oven.
6. Nitrile gloves.

3.2 Procedures

1. Weigh approximately 2.0000 g of ground sample into a Falcon tube of 50 mL and record the sample weight (*see* **Notes 1–3**).
2. Take the tubes with samples and caps to a connected fume hood, proceed to add exactly 8 mL of chloroform, 16 mL of methanol, and 6.4 mL of distilled water, and cap the tubes (*see* **Notes 4 and 5**).
3. Place the tubes on a tube shaker for 30 min, after stop shaking.
4. In the tubes, using a graduated pipettes, add exactly 8 mL of chloroform and 8 mL of 1.5% sodium sulfate solution, cap, and shake for another 2 min.

5. Allow the layers to separate naturally or centrifuge at 1500 rpm for 2 min.

Typically, after centrifugation, the Falcon tube presents with three layers:

- Upper: methanol and water.
- Middle: sample mixed with water and solvent.
- Lower: chloroform with extracted fat.

6. In the tubes, insert the graduated pipette, withdraw approximately 10–12 mL with a 20 mL graduated pipette from the bottom layer (chloroform) and place them in a Falcon tube (capacity 15 mL) containing approximately 1 g of anhydrous sodium sulfate, cap, and shake.
7. Filter using a filter paper and collect the filtrate in a 20 mL beaker.
8. Measure, with a volumetric pipette, exactly 5 mL of the filtrate and pour it into a previously weighed beaker of 10 mL.
9. Place the beaker of 10 mL in the fume hood until the chloroform evaporates.
10. Place the beaker of 10 mL after chloroform evaporation in an oven at 105°C for 1 h until all the solvent is evaporated and only the fat remains in the beaker.
11. Cool in a desiccator and weigh (beaker + lipid).
12. Calculate the lipid content of the sample as described in Sect. 3.

4 Results

To calculate the results, it is important to remember that during the analysis, 8 mL of chloroform was added in 2 stages, totaling 16 mL of chloroform. In other words, the total fat content of the sample is contained in 16 mL of chloroform. However, at the end of the experiment, only 5 mL of the filtrate was taken for evaporation in the oven. Therefore, during the calculations, it is necessary to convert the proportions before calculating the lipid content, according to Eqs. (1)–(3).

$$\text{Lipids in 5 mL of filtrate (g)} = (\text{BW} + \text{lipid}) - (\text{Empty beaker weight}) \quad (1)$$

BW: weight of the beaker (g)

$$\text{Lipids in 16 mL of chloroform (g)} = \frac{\text{Lipids in 5 mL of filtrate} \times 16}{5} \quad (2)$$

Table 1

Spreadsheet of records for collected data in the analysis

Lipid determination (%)											
Sample											
Analyst											
Date											
Sample identification (code and reference)	Number	Replicate	Sample weight in the Tube (g)	Empty beaker weight (g)	Beaker weight + Fat (g)	Lipids in 5 mL of chloroform (g)	Lipids in 16 mL of chloroform (g)	Total lipids %	Average (%)	SD	CV
	1	CS1									
CS*	2	CS2									
	3	CS3									

*Example for recording data of a starch sample. *CS*, corn starch; *SD*, standard deviation; *CV*, coefficient of variation

Table 2
Example of results for determining lipids in starches

Lipid determination (%)											
Sample											
Analyst											
Date											
Sample identification (code and reference)	**Number**	**Replicate**	**Sample weight in the Tube (g)**	**Empty beaker weight (g)**	**Beaker weight + Fat (g)**	**Lipids in 5 mL of chloroform (g)**	**Lipids in 16 mL of chloroform (g)**	**Total lipids %**	**Average (%)**	**SD**	**CV**
	1	CS1	2.0025	35.1646	35.1654	0.0008	0.0026	0.1278			
CS*	2	CS2	2.0026	35.1138	35.1146	0.0008	0.0026	0.1278	0.1331	0.0092	6.8968
	3	CS3	2.0036	35.3633	35.3642	0.0009	0.0029	0.1437			

*Example for recording data of a starch sample. *CS*, corn starch; *SD*, standard deviation; *CV*, coefficient of variation

$$\text{Lipids }(\%\text{org}/100\text{ g}) = \frac{\text{Lipids in 16 mL of chloroform} \times 100}{\text{Sample weight (g)}} \quad (3)$$

After obtaining the moisture content of the three replicates per sample, calculate the average, standard deviation (SD), and coefficient of variation (C.V.), as provided by data tabulation programs such as Microsoft Excel or Google Sheets. The standard deviation is considered optimal when its value is as close to zero as possible. A coefficient of variation below 10% indicates a dispersion close to ideal for this type of analysis as described in Tables 1 and 2 in the **Notes**.

5 Notes

1. Before commencing the analysis, it is necessary to separate the correct glassware for use and prepare the beakers and Falcon tubes.
2. Tare the beakers of 10 mL by drying them in an oven with air circulation at 130°C for 1 h. Remove and place them in a desiccator to cool for up to 40 min. Weigh them on an analytical balance and record the weight (weight of the empty beaker).
3. Selecting Falcon tubes with caps, and performing the leak test to verify if there are any leaks, choose only the capped tubes without leaks. Wash them according to the procedure, dry them, and label them with the sample codes (tube and caps), to proceed with the analysis.
4. The starch sample's moisture content must be calculated, so the moisture value is deducted from the volume of water to be added, or a dry sample can be used. If a dry sample is used, the conversion mentioned below is not necessary.
5. For example: 2 g of sample with 10% moisture means we have 1.8 g of sample and 0.2 g of moisture. Since the density of water is 1, 0.2 g = 0.2 mL. Hence: 6.4 – 0.2 = 6.2 mL of water to be added.

Acknowledgments

The authors would like to thank the National Council for Scientific and Technological Development (CNPq), Brazil, for the scholarships awarded to MTP Clerici (312660/2023-5) and the São Paulo Research Foundation (FAPESP), Brazil (Process Number 2023/12726-8) for the postdoctoral fellowship to BL Tagliapietra.

Additionally, the authors express gratitude to the Coordination for the Improvement of Higher Education Personnel (CAPES, financial code 001), Brazil, for the scholarships awarded to NA Brandão (Process 88887.605710/2021-00), and for supporting the Graduate Program in Food Technology at the Universidade Estadual de Campinas (UNICAMP), São Paulo, Brazil.

References

1. Cecchi HM (2007) Theoretical and practical foundations in food analysis, 2nd edn. UNICAMP, Campinas, p 208
2. Oliveira LA (2010) Laboratory manual: physicochemical analyses of fruits and cassavas, 1st edn. Embrapa Cassava and Fruit, Cruz das Almas, p 248
3. Bligh EC, Dyer WJ (1959) A rapid method of total lipid. Extraction and purification. Can J Biochem Physiol 37(8):911–917
4. American Association of Cereal Chemists (2010) AACCI approved methods of analysis, 11th edn. AACC International, St. Paul
5. Association of Official Analytical Chemists (2006) AOAC, 18th edn. Association of Official Analytical Chemists, Washington, DC

Chapter 4

Protein

Natali Alcantara Brandão, Bruna Lago Tagliapietra and Maria Teresa Pedrosa Silva Clerici

Abstract

Proteins are polymers of amino acids linked to each other through peptide bonds and compounds of carbon, hydrogen, oxygen, nitrogen, and sulfur. They can be present in the membranes of the amyloplast, which stores starch. The basis of the process of determining proteins in food is the digestion of organic matter by heating with sulfuric acid until the nitrogen present in the sample is transformed into ammonium sulfate. Then, distillation is performed with concentrated sodium hydroxide and heating to release ammonia into a known volume of a boric acid solution, forming ammonium borate. The formed ammonium borate is titrated with a standardized acidic solution. Then the total nitrogen content is converted into the protein present in the sample.

Keywords Starch, Proteins, Nitrogen, Kjeldahl

List of abbreviations

AM_N	Atomic mass of nitrogen = 14 u
CF_{HCl}	Correction factor of the HCl solution
CF_{NaOH}	Correction factor of the 0.275 mol/L NaOH solution
CS	Corn starch
CV	Coefficient of variation
D_{HCl}	Density of HCl
DMC_{NaOH}	Desired molar concentration of the NaOH solution (g/mol)
m_{NaOH}	Mass of NaOH to be determined (g)
MC_{HCl}	Molar concentration of the HCl solution, in this case = 0.02 mol/L
MM_{HCl}	Molecular mass of HCl = 36.46 g/mol
MM_{NaOH}	Molecular mass of NaOH: 39.9964 g/mol
N	Content of nitrogen present in the sample (%)
$Purity_{HCl}$	Purity of the HCl used

Maria Teresa Pedrosa Silva Clerici and Pedro H. Campelo (eds.), *Basic Protocols in Starch Characterization*, Methods and Protocols in Food Science, https://doi.org/10.1007/978-1-0716-5116-2_4,

RMC_{HCl}	Real molar concentration of the HCl solution (mol/L)
RMC_{NaOH}	Real molar concentration of the NaOH solution (mol/L)
SD	Standard deviation
TMC_{HCl}	Theoretical molar concentration of the HCl solution used (in this case, 0.02 mol/L)
TMC_{NaOH}	Theoretical molar concentration (desired) of the NaOH solution, in this case, 0.275 mol/L
V	Volume (mL)
$V_{solution}$	Volume of solution to be prepared (L)

1 Introduction

Proteins are highly complex polymers composed of 20 different amino acids. These amino acids are linked through substituted amine bonds with partial double-bond character, emphasizing the unique structural property of protein polymers. The functional diversity of proteins fundamentally resides in the variety of three-dimensional shapes that can be produced by rearranging the amino acid sequence in proteins. These biopolymers contain about 50%–55% carbon, 6%–7% hydrogen, 20%–23% oxygen, 12%–19% nitrogen, and 0.2%–3% sulfur [1].

Among the various methods used for protein determination, the Kjeldahl protocol is the most commonly used, being simple and reliable, allowing for the indirect determination of proteins. The Kjeldahl method determines the total nitrogen of the food, mainly present in the → amino group (-NH_2) of proteins. However, during the analysis, nitrogen from other sources is also determined, such as nitrogenous bases, ammonia salts, and free amino acids [2]. The protein content indicates the degree of extraction and purification of starch, when derived from cereals, legumes, pseudocereals, it tends to have a higher protein content due to the presence of amyloplast membranes and endosperm protein residues. In tubers, which naturally have a lower protein content, the nitrogen content should be lower than that of cereals. In purified starches, it is expected that all N from soluble proteins and nitrates and nitrites will not be present.

Determining the protein content present in starch will be important in the following cases:

(a) *Extraction process*: if protein levels are above 2% in starches, they should be purified again.

(b) *Starch to be used for modified starch production*: in this case, protein is more reactive than starch, so modification may occur in the protein initially, leading to changes in starch properties. It will be difficult to distinguish and quantify how much starch has been modified. It is also dangerous because proteins can form reactive complexes and antinutrients, which will impair starch quality.

The method is basically divided into three steps: digestion, distillation, and titration.

1.1 Digestion of Organic Matter

Using a protein digestion block, organic nitrogen is converted into ammonium sulfate ($(NH_4)_2SO_4$) in a medium containing concentrated sulfuric acid, simultaneously decarboxylating and burning organic compounds that are converted into SO_2, CO_2, H_2O, among others, which leave through evaporation.

In this stage, catalysts are used, such as potassium sulfate (K_2SO_4), which increases the boiling point of sulfuric acid (H_2SO_4) from 180 to 400°C, due to the formation of S_2O_7, accelerating digestion, and copper sulfate ($CuSO_4$), which increases the oxidation power of oxygen.

$$\text{Organic matter with N} + H_2SO_4 \xrightarrow[H_2O]{CuSO_4 + K_2SO_4\Delta} (NH_4)_2SO_4 + SO_2 + CO_2 +$$

1.2 Distillation

This is the stage where ammonia $(NH4)_2SO_4$ is separated and collected in a receiving solution, using a micro-*Kjeldahl* nitrogen distiller.

The nitrogen captured during digestion in the form of ammonium sulfate reacts with sodium hydroxide (NaOH) in an exergonic reaction, resulting in the formation of ammonium hydroxide ($2NH_4OH$), an unstable compound that quickly transforms into ammonia (NH_3). The ammonia, in vapor form, is condensed and reacts with boric acid (H_3BO_3), resulting in the formation of ammonium acid borate ($NH_4H_2BO_3$).

$$(NH_4)_2SO_4 + 2NaOH \overset{\Delta}{\rightarrow} 2NH_4OH + Na_2SO_4$$
$$2NH_4OH \overset{\Delta}{\rightarrow} 2NH_3 + 2H_2O$$
$$2NH_3 + 2H_3BO_4 \rightarrow NH_4H_2BO_3$$

1.3 Titration

This involves the quantitative determination of the ammonium contained in the receiving solution. Ammonium acid borate is titrated with hydrochloric acid (HCl) until a color change occurs, resulting in the formation of ammonium chloride (NH_4Cl) and boric acid as final products.

$$NH_4H_2BO_3 + HCl \rightarrow NH_4Cl + H_3BO_4$$

With knowledge of the volume of HCl consumed, the total nitrogen content as well as the protein content present in the sample can be calculated. Theoretical consideration suggests that on average, every 100 g of protein contains 16% nitrogen. Therefore, for every gram of nitrogen, there are 6.25 g of protein, as shown in Example 1. Thus, the factor 6.25 multiplied by the percentage of total nitrogen in the sample corresponds to the percentage of protein.

Example 1:

$$100 \text{ g} \rightarrow 16 \text{ g Nitrogen}$$
$$x \text{ g} \rightarrow 1 \text{ g Nitrogen}$$
$$x = 6.25 \text{ (conversion factor from nitrogen to \% protein)}$$

Since the protein present in the starch should only be residual from the extraction process, the factor 6.25 can be used.

2 Material

2.1 Equipment

1. Analytical balance.
2. Balance with a minimum capacity of 500 g.
3. Fume hood.
4. Nitrogen distiller (micro-Kjeldahl apparatus).
5. Oven.
6. Protein digester.

2.2 Material

1. 1 L amber glass jar.
2. 1 L black plastic container.
3. 1 L glass beaker.
4. 1 L volumetric flask.
5. 150 mL Erlenmeyer flask.
6. 2 L glass beaker.
7. 2 L plastic beaker.
8. 25 mL volumetric pipette.
9. 5 mL volumetric pipette.
10. 50 mL glass beaker.
11. Appropriate digestion tubes (suggestion: Semimacro for 100 mL = 25 mm diameter × 250 mm length).
12. Basin or another container suitable for an ice bath.
13. Glass funnel.
14. Parchment paper.

15. Petri dish.
16. Plastic funnel.
17. Plastic spatula.
18. Spatula.
19. Spatula.
20. Standard Personal Protective Equipment for a chemical analysis laboratory.
21. Starch sample.
22. Thermal gloves.
23. Tongs.
24. Wash bottle (with distilled water).
25. Watch glass.

2.3 Reagents

1. Boric acid (H_3BO_3).
2. Bromocresol green indicator.
3. Copper sulfate ($CuSO_4$) P.A..
4. Distilled water free of CO_2 (recently boiled).
5. Hydrochloric acid (HCl) P.A..
6. Methyl red indicator.
7. Potassium biphthalate [C_6H_4(COOH)(COOK)].
8. Potassium sulfate (K_2SO_4) P.A..
9. Selenium.
10. Sodium hydroxide (NaOH) P.A..
11. Sulfuric acid P.A. ($d = 1.85$).
12. Sulfuric acid P.A. (H_2SO_4; $d = 1.85$ - 98%).

2.4 Solutions

1. 50% Sodium hydroxide (NaOH) solution.
2. 2% Boric acid (H_3BO_3) solution with mixed indicator.
3. Catalyst mixture (K_2SO_4 + $CuSO_4$, in a 2:1 ratio, respectively).
4. 0.275 mol/L NaOH solution standardized with Potassium biphthalate.
5. 0.02 mol/L Hydrochloric acid (HCl) solution standardized with 0.275 mol/L NaOH.

3 Method

Protein Fraction Determination by the Kjeldahl method (*see* **Note 1**) [3–5].

3.1 Preparation of Solutions

3.1.1 50% Sodium Hydroxide (NaOH) Solution

1. Prepare a container with ice and set it aside.
2. Add about 1.2 L of recently boiled and cooled distilled water into a 2 L plastic beaker.
3. Weigh 500 g of NaOH into a 2 L plastic beaker.
4. In a connected fume hood, place the NaOH beaker on an ice bath.
5. Slowly add the recently boiled and cooled distilled water (do not add all the water at once).
6. Using a plastic spatula, mix the NaOH solution with water until dissolved.
7. Keep the NaOH solution beaker immersed in the ice bath, in a connected fume hood, until the solution cools completely.
8. Transfer the cold NaOH solution using a plastic funnel to a 1 L volumetric flask.
9. Fill the volumetric flask to the mark (up to the 1 L meniscus), quickly remove the prepared solution from the glass flask, transferring it, with the aid of a plastic funnel, to a plastic container with a lid.
10. Label and store in an appropriate location.

3.1.2 2% Boric Acid (H_3BO_3) Solution with Mixed Indicator

1. Weigh 20 g of H_3BO_3 into a 50 mL beaker.
2. Transfer the H_3BO_3 to a 1 L beaker.
3. Rinse the 50 mL beaker to ensure no residue remains.
4. Add another 500 mL of distilled water and mix until dissolved.
5. Add to the solution: 5 mL of methyl red + 25 mL of bromocresol green.
6. Mix well and transfer the mixture to a 1 L volumetric flask and fill up the volume with distilled water.
7. Using a funnel, transfer the solution to an amber glass jar.
8. Label and store in an appropriate location.

3.2 Catalyst Mixture

1. The ratio of the mixture should be 2 (potassium sulfate): 1 (copper sulfate).
2. Weigh into a glass beaker.
3. 94 g of potassium sulfate—K_2SO_4.
4. 47 g of copper sulfate—$CuSO_4$.
5. 1 g of selenium.
6. Mix with a spatula and transfer to a glass jar.
7. Label and store in an appropriate location.

3.3 Sodium Hydroxide (NaOH) 0.275 mol/L Solution Standardized with Potassium Biphthalate

3.3.1 Preparation of 0.275 mol/L NaOH Solution

1. Calculate the mass of NaOH required to prepare 1 L of solution (Eq. 1);
2. According to the mass determined in Eq. (1), weigh 11 g of NaOH in a plastic beaker;
3. Place this beaker on an ice bath in a fume hood.
4. Add approximately 600 mL of distilled water free of CO_2.
5. Mix with a plastic spatula until dissolved.
6. Allow to cool to room temperature.
7. Transfer to a 1 L volumetric flask (which has already been rinsed with distilled water).
8. Remember to rinse the beaker with a very small amount of distilled water so that any remaining residue is transferred to the volumetric flask.
9. Adjust the volume with distilled water.
10. Using a funnel, transfer the solution to a plastic container.
11. Label and store in an appropriate location.

$$\mathrm{DMC}_{\mathrm{NaOH}} = \frac{m_{\mathrm{NaOH}}}{\mathrm{MM}_{\mathrm{NaOH}} \times V_{\mathrm{solution}}} \quad (1)$$

Where

- $\mathrm{DMC}_{\mathrm{NaOH}}$ = Desired molar concentration of the NaOH solution (g/mol);
- m_{NaOH} = Mass of NaOH to be determined (g);
- $\mathrm{MM}_{\mathrm{NaOH}}$ = Molecular mass of NaOH: 39.9964 g/mol;
- V_{solution} = Volume of solution to be prepared (L).

3.3.2 Standardization of 0.275 mol/L NaOH with Potassium Biphthalate

1. Weigh 2 g of potassium biphthalate onto a Petri dish.
2. Place the dish in an oven at 110°C for 2 h to dry.
3. Weigh 0.5 g of biphthalate on a watch glass, record the weight, and transfer to an Erlenmeyer flask (repeat this step 3 times for triplicate analysis).
4. Using a graduated cylinder, transfer 25 mL of water to each Erlenmeyer flask.
5. Add three drops of phenolphthalein to each solution.

Table 1
Spreadsheet for recording data from the standardization of the NaOH solution

Erlenmeyer n°	Weight of potassium biphthalate (g)	$Volume_{NaOH}$ (mL) titrated	RMC_{NaOH} (mol/L)	TMC_{NaOH} (mol/L)	CF_{NaOH}	Average CF_{NaOH}
1				0.275		
2				0.275		
3				0.275		

6. After setting up the environment, fill a burette with the 0.275 mol/L NaOH solution.
7. Titrate the potassium biphthalate solution with 0.275 mol/L NaOH until the color turns slightly pink.
8. Record the volume of NaOH used (repeat the procedure three times).
9. Collect and record the data in a spreadsheet as shown in Table 1.
10. Calculate the Real Concentration of the NaOH solution according to Eq. (2).
11. Calculate the Correction Factor (CF) of the NaOH solution, according to Eq. (3).
12. The Correction Factor should be used in the standardization calculations of the 0.02 mol/L HCl solution.

$$\mathrm{RMC}_{\mathrm{NaOH}} = \frac{m_{\mathrm{Potassium\ biphthalate}} \times 1000}{V_{\mathrm{NaOH}} \times \mathrm{MM}_{\mathrm{Potassium\ biphthalate}}} \tag{2}$$

Where

- $\mathrm{RMC}_{\mathrm{NaOH}}$ = Real molar concentration of the NaOH solution (mol/L);
- m = Mass of potassium biphthalate added to the Erlenmeyer flask (g);
- V_{NaOH} = Volume of NaOH used in titration (mL);
- $\mathrm{MM}_{\mathrm{Potassium\ Biphthalate}}$ = Molecular mass of potassium biphthalate ($C_8H_5KO_4$) = 204.22 g/mol.

$$CF_{NaOH} = \frac{RMC_{NaOH}}{TMC_{NaOH}} \tag{3}$$

Where

- CF_{NaOH} = Correction factor of the 0.275 mol/L NaOH solution;
- RMC_{NaOH} = Real molar concentration of the NaOH solution found in Eq. (3);
- TMC_{NaOH} = Theoretical molar concentration (desired) of the NaOH solution, in this case 0.275 mol/L.

3.4 Preparation of 0.02 mol/L HCl Solution Standardized with 0.275 mol/L NaOH

3.4.1 Preparation of 0.02 N mol/L HCl Solution

1. Calculate the necessary volume of HCl to prepare the desired amount of solution with a concentration of 0.02 mol/L, according to Eq. (4).
2. From the volume found, prepare the solution according to the following items.
3. Using a graduated pipette, measure 1.61 mL of HCl.
4. Transfer to a 1 L volumetric flask.
5. Complete the volume with CO_2-free distilled water.
6. Transfer the solution to an amber glass bottle.
7. Label and store appropriately.

$$V_{HCl} = TMC_{HCl} \times MM_{HCl} \times V_{solution} \times \frac{1}{D_{HCl}} \times \frac{100}{purity_{HCl}} \quad (4)$$

Where

- V_{HCl} = Volume of HCl (to be determined) to prepare 1 L of solution (mL);
- TMC_{HCl} = Theorical molar concentration of the HCl solution, in this case = 0.02 mol/L;
- MM_{HCl} = Molecular mass of HCl = 36.46 g/mol;
- $V_{solution}$ = Volume of solution you will prepare; e.g., 1 L;
- D_{HCl} = Density of HCl (Check on the label. e.g., 1.19);
- $Purity_{HCl}$ = Purity of the HCl used (Check the label. e.g., 38%).

3.4.2 Standardization of 0.02 mol/L HCl Solution with 0.275 mol/L NaOH

1. Transfer, with a volumetric pipette, 25 mL of 0.02 mol/L HCl solution to an Erlenmeyer flask.
2. Add 2–3 drops of phenolphthalein.
3. In a buret add 0.275 mol/L NaOH solution.
4. Titrate the HCl solution until a slightly pink color appears.

Table 2
Spreadsheet for data collection and results of standardization of 0.02 mol/L HCl

Erlenmeyer n°	V_{HCl} (mL)	V_{NaOH} (mL)	TMC_{NaOH}	CF_{NaOH}	RMC_{HCl} (mol/L)	TMC_{HCl} (mol/L)	CF_{HCl}	Average CF_{HCl}
1	25 mL		0.275			0.02		
2	25 mL		0.275			0.02		
3	25 mL		0.275			0.02		

5. Note the volume of NaOH used.
6. Repeat the procedure three times.
7. Write down the results in a spreadsheet as per the example in Table 2.
8. Based on Eqs. (5) and (6), calculate the Correction Factor for the HCl solution.
9. Note down the average FC of the 0.02 mol/L HCl that will be used in the protein determination calculations.

$$RC_{HCl} \times V_{HCl} = (TMC_{NaOH} \times CF_{NaOH}) \times V_{NaOH} \quad (5)$$

Where

- RMC_{HCl} = Real concentration of the HCl you want to find (mol/L);
- V_{HCl} = Volume of the HCl aliquot transferred to the Erlenmeyer flask (mL);
- TMC_{NaOH} = Theoretical molar concentration of NaOH (in this case, 0.275 mol/L);
- CF_{NaOH} = (Eq. 4);
- V_{NaOH} = Volume of NaOH consumed in the titration (mL).

$$CF_{HCl} = \frac{RMC_{HCl}}{TMC_{HCl}} \quad (6)$$

Where

- CF_{HCl} = Correction factor of the HCl solution;
- RMC_{HCl} = Real Concentration of the HCl (mol/L) (Eq. 5);
- TMC_{HCl} = Theoretical molar concentration of the HCl (in this case, 0.02 mol/L);

3.5 Procedure for Protein Determination

3.5.1 Digestion

Prepare the tubes according to the number of samples and also to make the blank, and all analyzes must be in triplicate or more replicates. Number each tube with a heat-resistant pen and make sure they have a lid to be used during digestion. The blank tubes will go through all analysis steps except sample addition.

1. Quantitatively weigh 0.1–0.2 g of the sample in the digestion microtube (weight should not exceed) and record the weight (when performing calculations, convert from grams to milligrams).
2. The sample should be dried and degreased to prevent foaming during digestion. However, in the case of starches where the percentage of lipids is low, degreasing is not necessary, but the moisture content is very important and must be registered
3. Add 1 g of the catalytic mixture to the tube (Item 3.2).
4. Add 5 mL of H_2SO_4.
5. Place the tubes in the Kjeldahl digestion block with the fume hood on.
6. Heat initially from 50 to 100°C, increase the temperature by 50°C every 15 min, until reaching 350 to 400°C, always observing the behavior of the sample according to its composition.
7. The sample should be digested for 8 h or until it becomes greenish-blue transparent (almost colorless).

After digestion, turn off the digestion block and let the samples cool completely.

3.5.2 Distillation

Distillation should be performed in specific equipment as shown in Fig. 1.

1. Thirty minutes before turning on the distiller, turn on the water circulation device.
2. Add 10 mL of distilled water to each tube.
3. Start the distillation process using a blank tube.
4. Adapt the tube to the micro-Kjeldahl distillation apparatus.
5. Add 10 mL of the mixed indicator boric acid solution (Item 3.1.2) to a 250 mL Erlenmeyer flask.
6. Fit the Erlenmeyer flask into the distiller, in a position to receive the distillate, ensuring that the tip of the distiller is submerged in the solution.

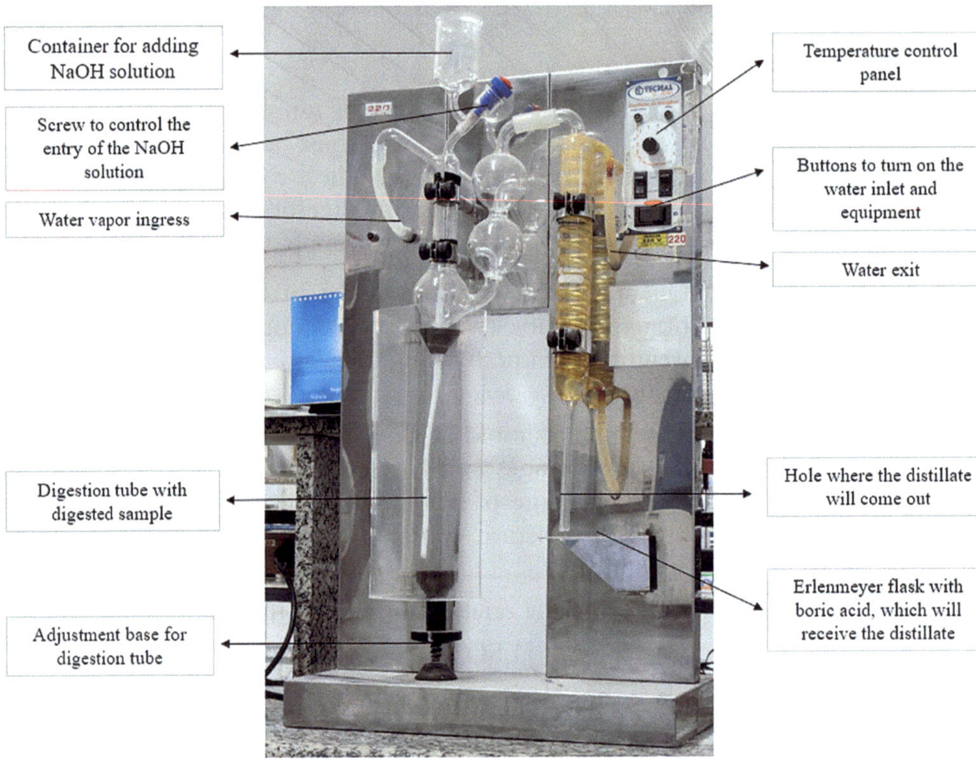

Fig. 1 Nitrogen distiller. Source: Authors' collection, 2025

7. Add 15–20 mL of the 50% NaOH solution to the specific inlet at the top of the distiller.
8. The distillate will be collected in the Erlenmeyer flask with boric acid. Approximately, 50 mL of the distillate should be collected. The color will change from pink to bluish-green (*see* **Note 2**).

3.5.3 Titration

1. In a buret, add the standardized 0.02 mol/L HCl solution (Item 3.4.1).
2. Titrate the collected distillate in the Erlenmeyer flask with boric acid and mixed indicator until a pink color appears.
3. Record the volume of 0.02 mol/L HCl solution used in the titration.
4. Record the results in a spreadsheet as shown in Tables 3 and 4 in **Notes**.

3.6 Results

1. Ensure that all units used are in the same standard. Use L/g or mL/mg.
2. If the sample was weighed in grams, convert to milligrams before starting the calculations.
3. Calculate the average volume of 0.02 mol/L HCl solution used to titrate the blanks.
4. Subtract the volume of HCl spent in the titration of each sample replica by the average volume spent on the blank, according to Eq. (7), thus obtaining the corrected HCl volume ($V_{\text{HCl corrected}}$);
5. Calculate the amount of nitrogen present in the sample according to Eq. (8);

$$V_{\text{HCl corrected}} = V_{\text{HCl sample}} - \text{Average } V_{\text{HCl bla.nk}} \quad (7)$$

$$N\ (\%) = \frac{V_{HCl\ corrected} \times TMC_{HCl} \times CF_{HCl} \times AM_N}{Sample\ weight} \quad (8)$$

Where

- N = Content of nitrogen present in the sample (%);
- V_{HCl} corrected = Volume of HCl used in the titration (mL), until color change, and corrected for white;
- TMC_{HCl} = Theoretical molar concentration of the HCl solution used (in this case, 0.02 mol/L);
- CF_{HCl} = Correction factor of the HCl solution (determined in item 4.5.2);
- AM_N = Atomic mass of nitrogen = 14 u;
- Sample weight = Initial weight of the sample (mg).

6. Calculate the amount of protein present in the sample according to Eq. (9).

$$\text{Protein } (\%) = N(\%)^{*}\ CF_{\text{starch}} \quad (9)$$

Where

- CF_{starch}= Protein conversion factor to starch: 6.25.

Table 3
Spreadsheet of data records collected in the analysis

Protein determination																	
Sample																	
Analyst																	
Location and Date																	
Sample identification (code and reference)	Number	Replicate	Sample weight (g)	Sample weight (mg) (*1000)	V_{HCl} sample (mL)	V_{HCl} blank (mL)	Average V_{HCl} blank (mL)	$V_{HCl\ corrected}$ (mL)	TMC_{HCl} (mol/L)	CF_{HCl}	AM_N	N (%)	CF Starch	Protein (%)	Average (%)	SD	CV
CS*	1	CS1				–	–		0.02		14		6.25				
	2	CS2				–	–		0.02		14		6.25				
	3	CS3				–	–		0.02		14		6.25				
Blank	4	B1			–			–	–	–	–	–	–	–	–	–	–
	5	B2			–			–	–	–	–	–	–	–			

*Example for recording data of a starch sample. *CS*, corn starch; *SD*, standard deviation; *CV*, coefficient of variation

Table 4

Example of results for determining protein in starches

Protein determination																
Sample																
Analyst																
Location and date																
Sample identification (code and reference)	**Number**	**Replicate**	**Sample weight (g)**	**Sample weight (mg) (*1000)**	**V_{HCl} sample (mL)**	**V_{HCl} blank (mL)**	**Average V_{HCl} blank (mL)**	**$V_{HCl\ corrected}$ (mL)**	**TMC_{HCl} (mol/L)**	**CF_{HCl}**	**AM_N**	**N (%)**	**CF Starch**	**Protein (%)**	**Average (%)**	**SD**	**CV**
CS*	1	CS1	0.2005	200.5000	0.2100	–	–	0.1600	0.02	0.9786	14	0.0219	6.25	0.1367	0.1395	0.0049	3.5202
	2	CS2	0.2004	200.4000	0.2100	–	–	0.1600	0.02	0.9786	14	0.0219	6.25	0.1367			
	3	CS3	0.2005	200.5000	0.2200	–	–	0.1700	0.02	0.9786	14	0.0232	6.25	0.1452			
Blank	4	B1	0.0000	0.0000	–	0.0500	0.0500	–	–	–	–	–	–	–	–	–	–
	5	B2	0.0000	0.0000	–	0.0500		–	–	–	–	–	–	–			

*Example for recording data of a starch sample. *CS*, corn starch; *SD*, standard deviation; *CV*, coefficient of variation

4 Notes

1. In this chapter, in addition to the description of the protein determination method, the preparation of solutions is also included, which is extremely important to obtain more precise results at the end of the analysis.
2. If using a different indicator, there may be differences in color tone.

Acknowledgments

The authors would like to thank the National Council for Scientific and Technological Development (CNPq), Brazil, for the scholarships awarded to MTP Clerici (312660/2023-5) and the São Paulo Research Foundation (FAPESP), Brazil (Process Number 2023/12726-8) for the postdoctoral fellowship to BL Tagliapietra. Additionally, the authors express gratitude to the Coordination for the Improvement of Higher Education Personnel (CAPES, financial code 001), Brazil, for the scholarships awarded to NA Brandão (Process 88887.605710/2021-00), and for supporting the Graduate Program in Food Technology at the Universidade Estadual de Campinas (UNICAMP), São Paulo, Brazil.

References

1. Damodaran S, Fenema OR (2017) Food chemistry. Boca Raton, CRC Press, 1107 p
2. Boas EVB (1999) Nutritional evaluation of foods. UFLA/FAEPE, Lavras, p 53
3. American Association of Cereal Chemists (2010) AACCI approved methods of analysis, 11th edn. AACC International, St. Paul
4. Association of Official Analytical Chemists (2006) AOAC, 18th edn. Association of Official Analytical Chemists, Washington, DC
5. Instituto Adolfo Lutz (2008) IAL—Analytical Standards of the Adolfo Lutz Institute. Physicochemical methods for food analysis, 4th edn. São Paulo, São Paulo, p 1020

Check for updates

Chapter 5

Total Dietary Fiber

Natali Alcantara Brandão, Bruna Lago Tagliapietra, Mária Herminia Ferrari Felisberto and Maria Teresa Pedrosa Silva Clerici

Abstract

Dietary fibers are composed of a combination of various polysaccharides that perform distinct functions in the body, being classified as soluble and insoluble based on their solubility in water. In starches, fibers may be present in minimal amounts, below 0.5%, as during the extraction process, those that are insoluble are the ones that are most easily extracted, whereas the soluble ones are difficult to remove. Enzymatic–gravimetric methods are the most recommended for the analysis of dietary fibers in foods, including starches.

Keywords Carbohydrates, Enzymatic–gravimetric method, Insoluble fibers, Soluble fibers, Starch

List of abbreviations

AA (g)	Average Ash (g) - Eq. (7).
AAB	Average ash of blank (g).
ACPB	Average of the corrected protein in the blank (g)—Eq. (6).
ACPS (g)	Average of the corrected protein sample (g)—Eq. (6).
$AM_N/100$ (u)	Atomic mass of nitrogen/100 (0.014 u) (protein content calculation).
ATBR (g)	Average TBR (g).
ATDF	Average TDF (g)
C	Celite
C_f	Fixed celite
CF_{starch}	Conversion factor of nitrogen into protein in the evaluated sample (starch): 6.25.
CF_{HCl}	Correction factor of the HCl solution (Which must be determined beforehand as described in Chapter 3 of Proteins of this book).

Maria Teresa Pedrosa Silva Clerici and Pedro H. Campelo (eds.), *Basic Protocols in Starch Characterization*, Methods and Protocols in Food Science, https://doi.org/10.1007/978-1-0716-5116-2_5,

CP (g)	Corrected protein content in relation to TDR (g).
CV	Coefficient of variation.
D_{HCl}	Density of HCl (Check on the label. e.g., 1.19)
DR	Dry residue
Eq.	Equation
MC_{HCl}	Molar concentration of the HCl solution (0.02 or 0.325 or 12 mol/L)
MM_{HCl}	Molecular mass of HCl = 36.46 g/mol
SD	Standard deviation.
TBR (g)	Total blank residue (g)—Eq. (8).
TDF (%)	Total dietary fiber (%)
TDR (g)	Total dry residue (g)
TDRB	TDR blank calculated according to Eq. (3)
TMC_{HCl} (mol/L)	Theoretical molar concentration of the HCl (0.02 mol/L) solution used in titration
V_{HCl} (mL)	Volume of HCl spent on protein titration (mL)
WC (g)	Weight of the empty crucible (g)
WCC (g)	WC + celite (g)
WCC_f (g)	WC + fixed celite (g) (After fixing the celite in the crucible using ethanol solution, and drying in the oven)
WCC_fA (g)	Weight of the crucible + fixed celite + ash (g) (crucible weight with the ashes remaining after burning the organic matter in the muffle furnace)
WCC_fDR (g)	WCC_f+ dry residue (g) (after drying the crucible with the digestion residue in the oven)
WS (g)	Weight sample (g) (initial weight of the sample added to the Erlenmeyer flask to start digestion)
WSP (g)	Weight sample protein (g) (weight of the dry residue added to the digestion tube for protein determination)

1 Introduction

Dietary fiber is not a single substance but rather a combination of various interconnected polysaccharides, such as cellulose, hemicellulose, β-glucans, pectins, gums, and mucilages [1], including components of plant cell walls that are not digested, such as lignin, resistant starch, inulin, oligofructose, and chitosans. In starches, fiber will be present in quantities below 0.5%, but in resistant starches, the fiber content should be close to the label's starch content. When starches are modified by cross-linking or to increase resistance to digestion, this method can be used to evaluate the effectiveness of the procedure. Since fiber content must be declared on food labels, many researchers prefer to use this method instead of specific methods to determine resistant starch content. Methods for determining dietary fibers can be classified into:

a. *Non-enzymatic-gravimetric methods* → Determination of crude fibers in neutral and acid detergents involves removing soluble fractions present in foods, focusing on the analysis of insoluble residue. The limitation of this method is that it does not quantify soluble fibers, such as pectins.

b. *Enzymatic–gravimetric methods* → This is the most appropriate method for fiber analysis in foods and uses enzymes that simulate an in vitro digestive process until only the undigested fractions remain, including fibers and resistant proteins. This more comprehensive approach allows for a more accurate and inclusive assessment of the different forms of fiber present in starch. Currently, there are several methods to evaluate dietary fiber content, all of which involve using enzymes to eliminate free sugars, starch, and proteins. Once enzymatic hydrolysis is complete, quantification can be performed gravimetrically.

2 Material

2.1 Equipment

1. Alcoholometer.
2. An analytical balance capable of weighing up to 0.1 mg.
3. Balance that weighs over 500 g.
4. Fume hood.
5. Desiccator.
6. Homogenizer.
7. Incubator.
8. Muffle furnace;
9. Nitrogen distiller (micro-Kjeldahl apparatus).
10. pH meter, calibrated to pH 4.0 and pH 7.0.
11. Protein digester.
12. Water bath (60–65°C and 95–100°C).
13. Vacuum pump with hoses and Buchner funnel.

2.2 Materials

1. 100 μL micropipette (precalibrated).
2. 150 mL Erlenmeyer flasks (two per sample + two for blank).
3. 1 L amber glass container with lid.
4. 1 L black plastic container with lid.
5. 1 L volumetric flask.
6. 2 L plastic beaker.
7. 35 mesh sieve.

8. 5, 20, and 25 mL volumetric pipettes (wash and dry at room temperature).
9. 500 mL Erlenmeyer flasks (four per sample + four for blank).
10. 50 mL, 100 mL, and 2 L glass beakers.
11. Aluminum foil to cover Erlenmeyer flasks.
12. Appropriate digestion tubes, semi-macro recommended for 100 mL = 25 mm diameter ×x 250 mm length (two per sample + one for blank).
13. Filter crucible (Gooch) with sintered glass filter plate (50 mL) of #2 porosity or 40–60 ASTM (four per sample + four for blank).
14. Tongs.
15. Glass beads.
16. Parchment paper.
17. Pasteur pipettes.
18. Petri dish.
19. Plastic and glass funnels.
20. Plastic and glass spatulas.
21. Plastic tray for ice bath.
22. Sample of starch, dried and with a particle size below 32 mesh.
23. Standard laboratory individual protection equipment for chemical analysis.
24. Thermal gloves.
25. Wash bottle (with distilled water).
26. Watch glass.

2.3 Reagents

1. Acetone P.A.
2. Anhydrous dibasic sodium phosphate (Disodium phosphate) (Na_2HPO_4) P.A.
3. Boric acid (H_3BO_3) P.A.
4. Bromocresol green indicator.
5. Celite (filtration aid composed of: diatomaceous earth with 87.5% SiO_2, 1.0% Al_2O_3; 6.6% CaO; 0.4% Fe_2O_3 and 1.5 % $Na_2O + K_2O$).
6. Copper sulfate ($CuSO_4$).
7. Distilled water.
8. EDTA dihydrate (ethylenediaminetetraacetic acid).
9. Ethanol (CH_3CH_2OH) absolute P.A.
10. Hydrochloric acid (HCl) P.A.

11. Megazyme K-TDFR Kit (Megazyme International, Wicklow, Ireland), containing the three enzymes: α-amylase, protease, and amyloglucosidase (*see* **Notes 1 and 2**).
12. Methyl red indicator.
13. Monobasic sodium phosphate monohydrate ($NaH_2PO_4.H_2O$) P.A.
14. Oxalic acid dihydrate ($C_2H_2O4.2H_2O$) P.A.
15. Potassium hydrogen phthalate [C_6H_4(COOH)(COOK)].
16. Potassium sulfate (K_2SO_4).
17. Potassium hydroxide (KOH);
18. Sulfuric acid P.A. (H_2SO_4; $d = 1.85\%–98\%$).
19. Sodium hydroxide (NaOH) P.A.
20. Selenium.
21. Trisodium phosphate (Na_3PO_4) P.A.

2.4 Solutions

2.4.1 Solutions for Fiber Determination

1. 95% Ethanol.
2. 78% Ethanol.
3. Demineralizing solution.
4. Cleaning solution.
5. 0.08 mol/L Phosphate buffer (pH 6.0).
6. Standardized 0.275 mol/L Sodium hydroxide (NaOH) solution.
7. Standardized 0.325 mol/L Hydrochloric acid (HCl) solution.
8. Standardized 12 mol/L Hydrochloric acid (HCl) solution.

2.4.2 Solutions for Determining the Protein Content Resulting from the Fiber Fraction

1. Standardized 0.275 mol/L sodium hydroxide (NaOH) solution.
2. Standardized 0.02 mol/L hydrochloric acid (HCl) solution.
3. 50% Sodium hydroxide (NaOH) solution.
4. 2% Boric acid (H_3BO_3) solution with mixed indicator.
5. Catalyst mixture (K_2SO_4 + $CuSO_4$ 2:1).

The solutions, glassware, and equipment for determining the remaining protein content of the fiber residue should be prepared as described in Chapter 3 of Proteins of this book.

3 Method

There are some kits for dietary fiber determination that are commercially available and provide all the necessary enzymes for the analysis. These assay kits are designed to measure the total amount of dietary fiber in food samples and operate on the principles of the enzymatic-gravimetric method. Below, we will describe the enzymatic-gravimetric method, used with the Megazyme K-TDFR Kit (Megazyme International, Wicklow, Ireland), based on method no. 985.29 [2]. The enzymes can also be purchased separately and prepared as respective enzymatic solutions, but care must be taken to use the same type of enzyme to maintain the same conditions and to verify concentrations in international units (IU) for dilution adjustments. It is not advisable to use enzymes with pH, temperature, and agitation conditions different from those described in the methodology, but if this occurs, modifications to the method must be validated using the official AOAC method [2], and the entire procedure must be described in the methodology of the study and published as a modified method [3–7].

3.1 Procedure

3.1.1 Analysis Preparation

Several preliminary steps should be taken:

(a) Ensure that all necessary materials are available in the required quantities for the analysis.

(b) Samples of starch should have their moisture content known, and this information should be noted to correct the values to a dry basis in the calculations.

(c) It is not possible to interrupt the analysis for the preparation of new solutions because, when conducting the blank (in quadruplicate), it must strictly follow all proposed solutions and steps.

(d) At the end of the procedure, the protein analysis in the fibers should be carried out because it will represent only the residue of enzymes and proteins remaining in the sample fiber. Therefore, all materials and solutions necessary for protein determination should be ready for use.

3.2 Determination of the Amount of Solutions, Reagents, and Materials that will be used

Before starting to prepare the solutions, it is necessary to calculate the quantity that will be used, according to the number of samples that will be evaluated. Predetermining the amount of reagents needed aims to prevent a lack of materials during the analysis and also to avoid waste. Therefore, it is important to read the protocol carefully to calculate the quantities. Always consider that four replicates are needed for each sample, and four more blanks must be added to each test. Remember that of the four replicates, two will be used to determine the protein content present in the residue, and two will be used to determine the ash. The same will happen with blank.

3.3 Solution Preparation

3.3.1 Standardized 0.275 mol/L NaOH Solutions with Potassium Biphthalate

The 0.275 mol/L NaOH solution will be used for both pH adjustment during fiber digestion steps and for standardizing HCl solutions. The preparation and standardization are described in Sect. 4.4 described in Chapter 3 of Proteins of this book.

3.3.2 Standardized 0.02, 0.325, and 12 mol/L HCl Solutions

1. The preparation of the three HCl solutions in different concentrations should follow the same dilution and standardization principle.
2. Calculate the required volume of HCl to prepare the desired amount of 0.02, 0.325, or 12 mol/L concentration solution according to Eq. (1).
3. Prepare the solution based on the volume found.
4. Using a graduated pipette, measure the V (L) volume of HCl determined in Eq. (1).
5. Transfer to a 1 L volumetric flask.
6. Top up the volume with CO_2-free distilled water.
7. Transfer the solution to an amber glass bottle.
8. Label, date, and store properly at room temperature.
9. After preparation, the three obtained solutions must be standardized with 0.275 mol/L NaOH as described in Sect. 4.5.2 in Chapter 3 of Proteins of this book.

$$V_{\mathrm{HCl}} = \mathrm{CM}_{\mathrm{HCl}} \times \mathrm{MM}_{\mathrm{HCl}} \times V_{\mathrm{solution}} \times \frac{1}{D_{\mathrm{HCl}}} \times \frac{100}{\mathrm{Purity}_{\mathrm{HCl}}} \tag{1}$$

Where

- V_{HCl} = Volume of HCl (to be determined) to prepare 1 L of solution (mL);
- $\mathrm{MC}_{\mathrm{HCl}}$ = Molar concentration of the HCl solution (0.02 or 0.325 or 12 mol/L);
- $\mathrm{MM}_{\mathrm{HCl}}$ = Molecular mass of HCl = 36.46 g/mol;
- V(solution) = Volume of solution you will prepare; Ex.: 1 L;
- D_{HCl} = Density of HCl (Check on the label. Ex.: 1.19);
- Purity = Purity of the HCl used (Check the label. Ex.: 38%).

3.3.3 95% and 78% Ethanol Solutions

1. Both 95% and 78% Ethanol solutions should be prepared in the same manner, following the steps below:

2. Calculate the required volume of Ethanol for the preparation of each solution according to Eq. (2).
3. Measure the volume of ethanol determined by Eq. (2) and transfer it to a 1 L volumetric flask.
4. Top up the volume with distilled water.
5. Check the concentration using the alcoholometer.
6. Transfer to a glass container, label, date, and store at room temperature.

$$C_1 \times V_1 = C_2 \times V_2 \tag{2}$$

Where

- C_1: Concentration (purity) of absolute ethanol as described on the label (e.g., 99.5%);
- V_1: Volume of absolute ethanol needed;
- C_2: Concentration (purity) of the desired solution (e.g., 95% or 78%);
- V_2: Volume of solution to be prepared.

3.3.4 Demineralizing Solution for Cleaning Filter Crucibles

1. Dissolve 50 g of oxalic acid dihydrate in 700 mL of 95% Ethanol solution in a beaker.
2. After complete dissolution of the acid, add 50 mL of 12 mol/L HCl solution (prepared as described in Sect. 3.3.2).
3. Transfer the solution to a 1 L volumetric flask and top up the volume with 95% Ethanol solution (about 250 mL).
4. Transfer to a glass container, label, date, and store at room temperature.

3.3.5 Cleaning Solution for Crucibles

Prepare in an ice bath:

1. In a 1 L plastic beaker, add in sequence:
   - 200 g of potassium hydroxide (KOH)+
   - 50 g of trisodium phosphate+
   - 5 g of EDTA dihydrate (ethylenediaminetetraacetic acid).
2. Then dissolve the salts in about 500 mL of distilled water.
3. After complete dissolution, transfer the solution to a 1 L volumetric flask, and top up the volume with distilled water.
4. After completing the volume, do not agitate.
5. Transfer to a glass container, label, date, and store at room temperature.

3.3.6 0.08 mol/L pH 6.0 Phosphate Buffer Solution

1. Weigh 1.400 g of anhydrous disodium phosphate (Na_2HPO_4).
2. Weigh 9.68 g of monosodium phosphate monohydrate ($NaH_2PO_4.H_2O$).
3. Transfer to a beaker and add 700 mL of distilled water, and stir for 30 min using a magnetic stirrer to dissolve the salts.
4. Transfer to a 1 L volumetric flask and top up the volume with distilled water.
5. Allow to equilibrate for 24 h under refrigeration.
6. Adjust the pH to 6.0 using 0.275 mol/L NaOH or 0.325 mol/L HCl solutions.
7. Transfer the phosphate buffer solution to a glass container, label, date, and store under refrigeration.

3.4 Procedures

Carefully planning each step, preferably within the same week as the analysis, can ensure the success of the analysis, which is time-consuming and costly, after all materials are prepared (*see* **Note 3**).

3.4.1 Preparation of Celite and Crucibles

Cleaning and Preparation of Crucibles

The preparation of the crucibles should occur before the sample filtration step; hence, it is advisable to be the first step in executing the procedure.

1. Calculate the necessary number of Gooch crucibles to be used, which is 4 per evaluated sample +4 for the blank, as each sample is performed in quadruplicate, where 2 Gooch crucibles will be used to determine ash and 2 for protein contents.
2. Place the Gooch crucibles immersed in the cleaning solution for 1 h, which should be in a glass beaker.
3. Remove the Gooch crucibles with tweezers and place them in a beaker with distilled water. Wait about 30 min, change the distilled water three times, and remove the crucibles from the distilled water. If necessary, use a wash bottle with distilled water to remove residues of the cleaning solution adhered to the crucible.
4. Pass the demineralizing solution through the crucible using a vacuum filtration system, placing the crucible on the Büchner flask in two stages:
   (a) In the opposite direction of filtration (from bottom to top), filter 10 mL of the solution or until the solution passes smoothly and quickly;

Table 1
Spreadsheet for data collection and calculations to obtain the total dietary fiber (TDF) analysis result

Determination of total dietary fiber																						
Sample								Analyst											Date			
		Dry residue							Protein					Ash								
Sample	Replicate	WC (g)	WCC (g)	WCCf (g)	WS (g)	WCCfDR (g)	TDR (g)	WSP (g)	V HCl sample (mL)	V HCl blank (mL)	Protein %	CP (g)	ACPS (g)	WCCfA (g)	Ash (g)	AA (g)	TBR (g)	ATBR (g)	TDF %	ATDF %	SD	CV
CS*	1																					
	2																					
	3																					
	4																					
Blank protein	BP1																					
	BP2																					
Blank ash	BC1																					
	BC2																					

Predetermined values for protein calculations			
TMC_{HCl} (mol/L)	CF_{HCl}	$MA_N/100$	FC_{starch}
0.02			6.25

(b) Then place the crucible in the regular direction of filtration (from top to bottom) and filter another 10 mL of the solution.

5. Rinse the crucible with distilled water to remove any residue of the demineralizing solution adhered to it.
6. Pass 95% ethanol (20 mL) through the crucible using vacuum, in the normal direction of filtration.
7. Dry the crucibles in an oven at 105°C for 30 min and then incinerate in the muffle furnace at 525°C for 5 h, let cool and remove from the muffle when the temperature reaches 150°C.
8. Transfer the crucibles to a desiccator, let cool for 40 min, weigh, and record in Table 1 the Weight of the empty crucible (WC);
9. Keep the crucibles in a desiccator until the Celite is ready to be fixed.

Preparation and Fixation of Celite in Crucibles

1. Calculate the necessary amount of Celite to be used during the analysis (0.5 g of Celite per crucible);
2. Weigh the total amount of Celite on a Petri dish with a lid, in an oven at 105°C for 1 h. Drying should occur on the same day or at most the day before fixation in the crucibles.
3. After drying, keep the Celite inside the covered Petri dish in a desiccator containing desiccant (silica gel).
4. Weigh in the crucible (prepared in step 5.1.1.) 0.5000 g of Celite and record in Table 1 the WC + Celite (WCC); this weight will not directly interfere with the calculations but will serve for checks, if necessary.
5. Fix the Celite in the crucibles with 15 mL of 78% ethanol, allowing it to drain down the walls of the crucible while simultaneously rotating the container so that the bottom is completely filled. Finally, use a vacuum to remove the ethanol and form the Celite layer, avoiding the formation of holes in it.
6. Dry the crucibles in an oven at 105°C for 1 h, remove from the oven, transfer to the desiccator, and let cool for 40 min, weigh, and record in Table 1 the WC + fixed celite (WCC_f); If necessary, for verification, calculate the actual weight of fixed Celite.

3.4.2 Sample Preparation

1. Use a dry sample of starch. If the moisture content is higher than 10%, dry it in a vacuum oven (70°C) until constant weight is achieved as described in Chapter 1 of Moisture of this book (*see* **Note 4**).
2. Identify the Erlenmeyer flask with a pen designed for glassware, on the top (near the neck), so that during digestion in the water bath, it does not reach or remove the marking.

3. Grind the sample and use a sieve to standardize the particle size to 32 mesh (0.3–0.5 mm).
4. Weigh the ground and dried sample (0.5000 to 1.0000 ± 0.02 g) in a 500 mL Erlenmeyer flask and record the weight sample (WS) in Table 1.
5. Four replicates of each sample to be analyzed should be weighed to obtain four results of fiber residues, which will be directed two to protein and two to ash contents. Also, prepare four Erlenmeyer flasks without samples for blanks.

3.4.3 Sample Digestion

1. In a graduated cylinder, measure 50 mL of phosphate buffer (pH 6.0) and add it to each 500 mL Erlenmeyer flask already containing the sample.
2. Check the pH to be 6.0 ± 0.1 and adjust if necessary using 0.275 mol/L NaOH or 0.325 mol/L HCl solutions.
3. Using a micropipette, measure 50 µL of α-Amylase and add it to each Erlenmeyer flask. Cover the Erlenmeyer flasks separately with aluminum foil and place them in the water bath at 95–100°C.
4. Using a thermocouple, monitor the temperature until the α-Amylase solution with the sample in the Erlenmeyer flasks reaches 95–100°C to start the timing.
5. After 30 min, remove the Erlenmeyer flasks and let them cool slowly to room temperature (this may take a few hours).
6. At this point, it is advisable to adjust the bath temperature for the next use, which will be at 60°C.
7. Adjust the pH of the solution containing the samples in the Erlenmeyer flasks, cooled, to 7.5 ± 0.1 using 0.275 N NaOH solution.
8. After pH adjustment, measure 100 µL of protease with a micropipette and add it to the Erlenmeyer flasks, covering them again with aluminum foil.
9. Place them in the water bath, which by now should be adjusted to 60°C, place the Erlenmeyer flasks, and maintain continuous stirring at 70 rpm.
10. Using a thermocouple, monitor the temperature of the solution in the Erlenmeyer flask until it reaches 60°C, then start the timer and maintain it for 30 min, keeping the stirring.
11. Next, remove the Erlenmeyer flasks containing the solutions and cool them to room temperature.
12. After cooling, adjust the pH value of the solutions in the Erlenmeyer flasks to 4.5 ± 0.2 using 0.325 mol/L HCl solution.

13. Using a micropipette, measure 200 μL of amyloglucosidase solution and add it to the Erlenmeyer flasks containing the solutions. Cover each flask with aluminum foil.
14. Mix the solutions by gently shaking the Erlenmeyer flasks and transfer them to the water bath at 60°C with continuous stirring (70 rpm). Wait for the solutions to reach 60°C, start the timer, and keep it for 30 min, maintaining stirring.
15. Remove from the water bath and add 95% ethanol to the Erlenmeyer flasks, using a volume of ethanol corresponding to four times the volume of the solution in the Erlenmeyer flask (approximately 280 mL).
16. Cover the Erlenmeyer flasks with aluminum foil and let them stand at room temperature for 16 h or more, as soluble fiber precipitation will occur.
17. Moisten the Gooch crucibles, already prepared with Celite beforehand, with 5 mL of 78% ethanol to prevent Celite from detaching from the crucibles during filtration.
18. Using a vacuum filtration system, filter the solution from the Erlenmeyer flasks into the Gooch crucible with Celite. *Reminder:* Check the numbering of the crucible and the Erlenmeyer flask for each sample.
19. Maintain vacuum filtration, successively wash the residue retained in the Gooch crucible, following this sequence of solutions:
    (a) Measure 20 mL of 78% ethanol in a graduated cylinder, wash the sample with this solution, and repeat the process two more times.
    (b) Measure 10 mL of 95% ethanol in a graduated cylinder, wash the sample with this solution, and repeat the process once more.
    (c) Measure 10 mL of Acetone P.A. in a graduated cylinder, wash the sample with this solution, and repeat the process once more.
20. Transfer the Gooch crucibles containing the retained fibers to an oven at 105°C and keep them for 6 h or more.
21. Remove the crucibles from the oven, let them cool to room temperature in a desiccator for 30 min, weigh, and record in Table 1 the WCC_f + dried residue (WCC_f DR).
22. For each sample, separate two crucibles for protein analysis and two for ash analysis present in the residue.
23. Also separate four Erlenmeyer flasks for the analysis blank, two for protein content analysis and two for ash analysis (*see* **Note 5**).

3.4.4 Determination of Protein Content in the Sample Fibers

1. Separate two digestion tubes per sample + two tubes for the fiber blank and two for the specific protein blank.
2. Carefully remove the celite and the sample from the crucibles without scratching the filter part or the sides of it.
3. Grind all the material in a mortar to reduce the particle size.
4. Weigh the ground sample into the digestion tubes and note the weight sample protein (WSP) in Table 1; Repeat the same process for the remaining residue from the two crucibles of blank.
5. Do not add samples or residues to the specific protein blank tubes, only add the reagents as the following items;
6. In suitable tubes for protein digestion, add:
   a. About five glass beads (tip of the spatula).
   b. 1 g of catalytic mixture.
   c. 5 mL of sulfuric acid P.A. (perform the operation in a fume hood with the exhaust on).
7. Place the tubes containing the sample and reagents for digestion in the block at 450°C for 4 h (heat the block slowly and gradually due to the presence of celite: 50°C/30 min).
8. After digestion of the samples, proceed to distillation and titration according to the method described in Chapter 3 of Proteins of this book. Note in Table 1 the Volume of 0.02 mol/L HCl Solution used in the titration (V-HCl) of the distillate.

3.4.5 Determination of Ash Content in the Sample Fibers

1. For the ash determination, two crucibles of each sample will be used, in addition to two crucibles of the white sample.
2. Place the crucibles in the muffle furnace at 525°C for 5 h.
3. Remove the crucibles from the muffle furnace when it has cooled and reached between 130 and 180°C.
4. Transfer the crucibles to a desiccator, using tweezers, and allow them to cool to room temperature for 40 min.
5. Weigh the crucibles and note the WCC_f + ash (WCC_fA) in Table 1.
6. Remove the residue and clean the crucibles with suitable solutions for reuse.
7. For further details, described in Chapter 2 of Ash of this book.

3.5 Results

Fiber analysis is a protocol that demands great attention and care, both in execution and in handling the results. Therefore, all data must be collected with meticulous organization to ensure the most accurate results possible. To achieve this, we suggest a data collection spreadsheet, as exemplified in Table 2.

Table 2
Example of results for determining total dietary fiber in starches

Determination of total dietary fiber																						
Sample								Analyst											Date:			
		Dry Residue						Protein						Ash								
Sample	Replicate	WC (g)	WCC (g)	WCCf (g)	WS (g)	WCCfDR (g)	TDR (g)	WSP (g)	V HCl sample (mL)	V HCl blank (mL)	Protein %	CP (g)	ACPS (g)	WCCfA (g)	Ash (g)	AA (g)	TBR (g)	ATBR (g)	TDF %	ATDF %	SD	CV
CS*	1	40.0537	40.5564	40.5557	0.5025	40.5716	0.5179	0.5152	7.8500	0.00	1.4888	0.0077	0.0081	–	–	–	–	–	0.2146	0.2337	0.02	7.03
	2	49.0570	49.5595	49.5592	0.5036	49.5750	0.5180	0.5150	8.8000	0.00	1.6656	0.0086		–	–	–	–	–	0.2340			
	3	38.2507	38.7543	38.7557	0.5020	38.7688	0.5181	–	–	–	–	–	–	38.7476	0.4969	0.4976	–	–	0.2547			
	4	49.1278	49.6282	49.6327	0.5092	49.6458	0.5180	–	–	–	–	–	–	49.6262	0.4984		–	–	0.2314			
Blank protein	BP1	48.5567	49.0594	49.0626	0.0000	49.0579	0.5012	0.4035	6.2500	0.00	2.6526	0.0133	0,0119	–	–	–	0.0035	0.0084	–	–	–	–
	BP2	47.9584	48.4658	48.4655	0.0000	48.4682	0.5098	0.4196	5.0500	0.00	2.0611	0.0105		–	–	–	0.0121		–	–	–	–
Blank ash	BC1	48.7809	49.2854	49.2843	0.0000	49.2845	0.5036	–	–	–	–	–	–	49.2704	0,4895	0.4858	0.0059		–	–	–	–
	BC2	48.5227	49.0244	49.0259	0.0000	49.0326	0.5099	–	–	–	–	–	–	49.0048	0,4821		0.0122		–	–	–	–

Predetermined values for protein calculations			
TMC_{HCl} (mol/L)	CF_{HCl}	MA_N /100	FC_{starch}
0.02	0.9786	0.014	6.25

After completing the analysis and data collection, proceed to calculate the results. To facilitate understanding, these calculations have been divided into five stages: Dry Residue, Ash, Protein, Total Blank Residue and Total Dietary Fiber.

3.5.1 Total Dry Residue (TDR)

The dry residue consists of the sample that resisted digestive processes and was subsequently dried in an oven. This residue mainly comprises organic matter, including fibers and nitrogenous compounds (such as proteins), as well as inorganic matter (ashes), and celite. The weight of the dry residue will be decisive in all subsequent stages and should be calculated according to Eqs. (3) and (4).

$$\mathrm{TDR\,(g)} = \mathrm{WCC_fDR} - \mathrm{WC} \tag{3}$$

Where

- TDR = Total dry residue (g);
- WC = Weight of the empty crucible (g);
- $\mathrm{WCC_fDR} = \mathrm{WCC_f}$ + dry residue (g).

Repeating the same calculation for the blank crucibles, the weight of the Blank Residue (BR) was obtained.

3.5.2 Proteins

1. At this point, the results of the two replicates of each sample and also the residues of the two blanks will be calculated;
2. Before calculating the protein content, correct the volume of HCl solution used in titrating the sample and the blank residue. The correction must be made by subtracting the volume spent on the protein-specific blank titration, according to Eq. (4);
3. Repeat the same calculation (Eq. 4) for the results of the two blanks, thus determining the Protein of the Blank;
4. As the calculated protein content is related to the weight of dry residue added to the digestion tube, it is necessary to proportionally convert this content to the total amount of dry residue generated after oven drying (TDR) (where there is sample residue and celite) of according to Eq. (6), obtaining the corrected protein (CP);
5. Then calculate the average of the corrected protein of the two samples (ACPS);
6. Repeat the same calculations for the results of the two blanks, thus determining the average corrected protein for the blank (ACPB).

$$V_{\mathrm{HCl}}\ \mathrm{corrected} = V_{\mathrm{HCl}}\ \mathrm{sample} - V_{\mathrm{HCl}}\ \mathrm{blank} \tag{4}$$

To calculate the proteins, follow Eqs. (5) and (6).

$$\text{Protein }(\%) = \frac{V_{\text{HCl corrected}} \times \text{TMC}_{\text{HCl}} \times \text{CF}_{\text{HCl}} \times \left(\frac{\text{AM}_{\text{N}}}{100}\right) \times \text{CF}_{\text{starch}}}{\text{WSP}} \times 100 \quad (5)$$

Where

- V_{HCl} corrected = Volume of HCl solution used in titration (mL) (until color change) and corrected for blank volume;
- TMC_{HCl} = Theoretical molar concentration of HCl solution (in this case, 0.02 mol/L);
- CF_{HCl} = Correction factor of the HCl solution;
- $\text{AM}_{\text{N}}/100$ = Atomic mass of nitrogen/100 = 0.014 (u);
- $\text{CF}_{\text{starch}}$ = Conversion factor of nitrogen into protein: 6.25;
- WSP = Weight sample protein placed in the digestion tube (g).

$$\text{Corrected protein (g)} = \frac{\text{Protein }(\%) \times \text{TDR}}{100} \quad (6)$$

Where

- Corrected protein (CP): Amount of protein present in TDR (g);
- Protein (%): Protein content found in Eq. (4);
- TDR: Weight of the total dry residue calculated according to Eq. (2) (g).

3.5.3 P3-Ash

1. To calculate the ash content, follow Eq. (7).
2. Calculate the average amount of ash in the samples (AA).

$$\text{Ash (g)} = \text{WCC}_{\text{f}}\text{A} - \text{WC} \quad (7)$$

Where

- $\text{WCC}_{\text{f}}\text{A} = \text{WCC}_{\text{f}}$ + ash (g);
- WC = Weight of the empty crucible (g).

Repeat the same calculation for the results and average of the two blanks, thus determining the average ash of the blank (AAB);

3.5.4 Total Blank Residue

1. Calculate the total blank residue (TBR) according to Eq. (8).
2. Then calculate the average total blank residue (ATBR).

$$\text{TBR (g)} = \text{TDRB} - \text{ABP} - \text{ABA} \quad (8)$$

Where

- TBR = Total blank residue (g);
- TDRB = Total dry residue blank calculated according to Eq. (3);
- ACPB = Average of the corrected protein blank according to Eqs. (5) and (6);
- AAB = Average ash of blank according to Eq. (7).

3.5.5 Total Dietary Fiber (TDF)

1. Calculate the total dietary fiber present in the sample according to Eq. (9).
2. Calculate average TDF, standard deviation (SD) and coefficient of variation (CV) of the results.

$$\text{TDF}\ (\%) = \frac{\text{TDR} - \text{ACPS} - \text{AA} - \text{ATBR}}{\text{WS}} \times 100 \tag{9}$$

Where

- TDR = Total dry residue (g) (Eq. 3);
- ACPS = Average corrected protein sample (present in the total dry residue generated) according to Eqs. (5) and (6);
- AA = Average amount of ash (g), according to Eq. (7);
- ATBR = Average total blank residue (g);
- WS = Weight initial sample added to Erlenmeyer (g).

4 Notes

1. The α-amylase, protease, and amyloglucosidase enzymes should be taken out of the refrigerator at the time of use, and after use, they should be immediately stored under refrigeration.
2. It is recommended to transfer an approximate amount of the enzyme to be used to an Eppendorf tube to avoid the risk of contamination of the contents of the bottle during pipetting.
3. Verify if all equipment, such as the incubator, vacuum pump, digestion block, and distiller, is functioning properly. Pre-calibrate the balances and the digital potentiometer. Before starting the digestion of the samples, adjust the water level of the water bath so that it is sufficient to cover all the liquid in the Erlenmeyer flask.

4. Since the subject of this chapter is starches, degreasing is not necessary, as with other types of samples, as starches are expected to have minimal amounts of fat residues.
5. The residue remaining in the filtration flask of the vacuum filtration system should be disposed of in an appropriate waste disposal area for reagents.

Acknowledgments

The authors would like to thank the National Council for Scientific and Technological Development (CNPq), Brazil, for the scholarships awarded to MTP Clerici (312660/2023-5) and the São Paulo Research Foundation (FAPESP), Brazil (Process Number 2023/12726-8) for the postdoctoral fellowship to BL Tagliapietra. Additionally, the authors express gratitude to the Coordination for the Improvement of Higher Education Personnel (CAPES, financial code 001), Brazil, for the scholarships awarded to NA Brandão (Process 88887.605710/2021-00), and for supporting the Graduate Program in Food Technology at the Universidade Estadual de Campinas (UNICAMP), São Paulo, Brazil.

References

1. Mira GS, Graf H, Cândido LMB (2009) Retrospective view on dietary fibers with emphasis on beta-glucans in the treatment of diabetes. Braz J Pharm Sci 45(1):11–20
2. Association of Official Analytical Chemists (2005) AOAC approved methods of the American Association of Chemists, 18th edn. AOAC International, Gaithersburg, MD
3. American Association of Cereal Chemists (2010) AACCI approved methods of analysis, 11th edn. AACC International, St. Paul
4. Efimtseva EA, Chelpanova TI (2021) Пищевые волокна как модуляторы секреции гастроинтестинальных гормональных пептидов [Fibras alimentares como modulators of gastrointestinal hormonal peptide secretion] Вопросы питания 90(4), 20–35
5. Food Ingredients Brazil (2024) Dossier: dietary fibers. Number 3, 2008 https://revista-fi.com/upload_arquivos/201606/2016060604378001465822086.pdf. Accessed 29 Mar 2024
6. Silva CO (2016) Food science: principles of bromatology, 1st edn. Rubio, Rio de Janeiro, p 248
7. Soliman GA (2019) Dietary fiber, atherosclerosis, and cardiovascular disease. Nutrients 1(5):1155

Chapter 6

Starch Extraction by Aqueous Method

Hugo José Martins Carvalho, Bruna Lago Tagliapietra, Natali Alcantara Brandão and Maria Teresa Pedrosa Silva Clerici

Abstract

Laboratory extraction methods are of great importance for obtaining native starch in sufficient quantity and quality for characterization, especially from raw materials with little or no information in the literature. The aqueous extraction technique should be used initially, as it aims to obtain native starches with minimal modification, facilitating the separation of non-starch carbohydrates, fibers, sugars, proteins, lipids, ash, and other bioactive compounds such as pigments and antinutrients., This chapter discusses the aqueous extraction method, as well as essential precautions, standardization procedures, and strategies to optimize the yield and purity of the extracted starch.

Keywords Extraction, Aqueous, Yield, Starch

1 Introduction

Various extraction methods can be applied to obtain native starches, and they are already well established for cereal starches such as corn, wheat, and rice, as well as for tubers and roots like cassava and potato. However, for starches from other sources, such as seeds, forage grasses, and nonindustrialized tubers, initial extraction studies must be carried out with caution. The aim is to obtain purified starches with minimal modification in order to preserve their original physicochemical and technological properties.

This method can be applied to starches of different botanical origins, but it is essential to first determine their proximate composition and, if possible, perform qualitative (Lugol's test) or quantitative assessments of starch presence. These analyses help define the initial quantity of raw material required to achieve adequate starch yield for the intended laboratory analyses.

Maria Teresa Pedrosa Silva Clerici and Pedro H. Campelo (eds.), *Basic Protocols in Starch Characterization*, Methods and Protocols in Food Science, https://doi.org/10.1007/978-1-0716-5116-2_6,

Aqueous extraction is one of the methods that most closely preserves the native form of starch, as it avoids the use of chemical reagents. It is characterized by simplicity, sustainability, and efficiency across various botanical sources. The process generally includes soaking, wet grinding with water, sieving, and centrifugation, sufficient steps to isolate native starch. Additionally, this method maintains the chemical structure of the starch, allowing it to be studied in its original form and enabling the assessment of its properties and potential applications.

2 Material

1. Disposable gloves.
2. Glass beakers (50 mL, 600 mL, and 1 L).
3. Metal spatula.
4. Filter mesh (88 mesh).
5. 50 mL Falcon tubes with caps, previously weighed and labeled (tube and cap).
6. Distilled water wash bottle.
7. Petri dish.

2.1 Reagents

1. Freshly distilled deionized water.
2. Sodium hypochlorite (NaClO) is used exclusively for sample sanitization.

Aqueous starch extraction

2.2 Sample

Approximately 15 g of fresh or dry sample (grains, seeds, fruits, tubers, roots, or rhizomes). It needs to determine moisture as described in Chapter 1 to calculate the yield at the end of the process.

2.3 Equipment

1. Analytical balance with precision up to 0.0001 g, within calibration period and with available standard operating procedure.
2. Drying oven with air circulation, within calibration period and with available standard operating procedure.
3. Rotating centrifuge for 50 mL Falcon-type tubes with a cooling system, within calibration period and with available standard operating procedure.
4. Mixer or blender, with an available standard operating procedure.

3 Method

Aqueous starch extraction [1].

3.1 Procedure

3.1.1 Preanalysis Steps

1. Preselect the appropriate number of beakers according to the number of samples, and conduct analyses in triplicate. Label beakers according to sample identification numbers.
2. Prepare a data table in the laboratory notebook for recording results, as illustrated in Table 1 in the **Results**.

3.1.2 Starch Extraction

1. Begin by washing the sample under running water to remove dirt, stones, branches, or other debris (applicable to grains, seeds, fruits with peel, *Solanum lycocarpum*, banana, tubers, roots, and rhizomes).
2. Sanitize the samples using a 200 ppm sodium hypochlorite solution. The preparation for this concentration is described in Supplementary Material 1.
3. Immerse the samples in the sanitizing solution and allow them to rest for 15 min. Ensure that the sanitizer fully covers the samples and gently stir every 5 min during this period.
4. After the sanitization time, discard the sanitizing solution.
5. Rinse the samples thoroughly using freshly distilled deionized water to remove residual chlorine.
6. Repeat the rinsing process at least three times to ensure adequate chlorine removal.
7. If necessary, allow the samples to air dry overnight in a sieve or draining rack to remove remaining hypochlorite through evaporation.
8. Weigh the samples using an analytical balance (precision 0.01 g), using approximately 10–15 g, and record the exact weight.*

   **The entire process should be performed in triplicate.*
9. If required, peel the samples using a knife.
10. Transfer the weighed samples to a mixer, blender, or food processor.
11. Add freshly distilled water in a 1:3 ratio (sample:water) for processing.

Example:

For every 100 g of sample to be used:

12. 100 g of sample → ratio of 1

Thus

13. 1 → 100 g of sample

    3 → x g of water

Therefore

14. $x = 300$ g of distilled water
15. Process for 1–2 min at medium speed.
16. Transfer the mixture to a filtration mesh (88 mesh) for sieving.
17. Rinse the processing container using a wash bottle with small amounts of distilled water to recover any retained material, and proceed with filtration (*see* **Note 1**).
18. Transfer the filtrate to 50 mL Falcon tubes and seal with the corresponding caps.
19. Transport the tubes containing the starch to the centrifuge.*
    **Refer to Supplementary Materials 2 and 3 and Table 3 for detailed calibration and operational instructions for the centrifuge.*
20. Place tubes with similar weights in opposite positions in the rotor to ensure the center of gravity remains balanced.
21. Centrifuge at refrigeration temperature (~4 to 6°C) at 10,000 rpm for approximately 10 min.
22. After centrifugation, a separation of layers will be visible, with starch, due to its density settling at the bottom of the tube as illustrated in Fig. 1.
23. Discard the supernatant, and using a spatula, remove the upper layer (containing proteins/fibers/lipids).
24. Resuspend the sediment at the bottom of the tube in distilled water (*see* **Note 2**).
25. Repeat steps 18–20 until only starch remains in the tube and the supernatant water appears clear, free of coloration and suspended materials such as small fibers.
26. Resuspend the material in distilled water again and proceed to filtration using a Büchner funnel system connected to a vacuum pump.
27. Use a filter paper with a slightly smaller diameter than the base of the Büchner funnel to ensure a proper fit. High-quality filter papers are essential to prevent tearing and ensure effective filtration.
28. Place a preweighed filter paper (previously dried in an oven and labeled in pencil) into the Büchner funnel, ensuring it is evenly seated on the porous surface of the funnel.
29. Add a small amount of the solvent or solution to be filtered onto the filter paper. This helps secure the paper to the funnel, sealing the edges and preventing liquid and solids from passing underneath.

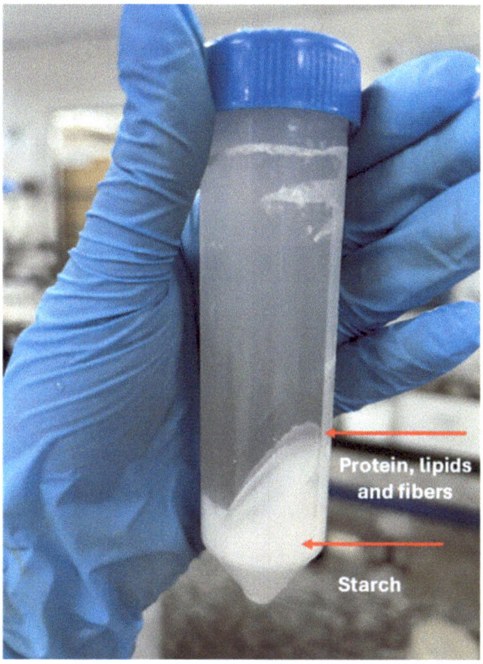

Fig. 1 Starch extraction with its separations after centrifugation, XE "Centrifugation" by difference in molecular weight. Source: Authors' collection, 2025

30. Begin applying vacuum to the filtration flask (Kitasato) connected to the Büchner funnel. The vacuum will draw the liquid through the filter paper, retaining the solids on the paper's surface (*see* **Note 3**).
31. Place the filter paper with the sample in a Petri dish and place it in the oven.
32. Dry in an air circulation oven at approximately 30–45°C with open air circulation to remove residual water overnight (~12 h) (*see* **Note 4**).
33. Place the plates in a desiccator containing silica gel (see "Moisture" chapter) and wait 40 min before weighing.
34. Weigh the starch obtained on an analytical balance.
35. Record all values on the data recording spreadsheet (Tables 1 and 2 in **Results**).

4 Results

The yield of extracted starch can be expressed by considering the amount of grain used initially, as well as the initial starch content contained in the raw material.

Table 1
Spreadsheet recording data collected during the analysis*

Starch yield determination based on initial sample (%)										
Sample										
Analyst										
Date										
Sample identification (code and reference)	**Number**	**Replica**	**Empty filter paper weight (g) (P1)**	**Initial sample weight (g) (P2)**	**1st Weighing crucible weight + dry starch (g) (P3)**	**Final starch weight (g) P3–P1**	**Yield (%)**	**Average (%)**	**D.P.**	**C.V.**
	1	CS1								
CS*	2	CS2								
	3	CS3								

*Example of data recording for a starch sample. *CS*, cassava starch; *SD*, standard deviation; *CV*, coefficient of variation

Table 2
Example of data collected in the analysis*

Starchyield determination based on initial sample (%)										
Sample	**Cassava starch (CS)**									
Analyst	**João**									
Date	**20/07/2025**									
Sample identification (code and reference)	**Number**	**Replica**	**Empty filter paper weight (g) (P1)**	**Initial Sample Weight (g) (P2)**	**1st Weighing crucible weight + dry starch (g) (P3)**	**Final starch weight (g) P3 - P1**	**Yield (%)**	**Average (%)**	**D.P.**	**C.V.**
	1	CS1	0.9986	14.5789	7.2596	66.2614	42.9456			
CS*	2	CS2	0.9998	14.6578	7.0159	6.0161	41.0437	42.2714	1.0650	2.5194
	3	CS3	0.9739	14.7825	7.3045	6.3306	42.8250			

*Example of data recording for a starch sample. *CS*, cassava starch; *SD*, standard deviation; *CV*, coefficient of variation

Calculate the total starch yield obtained according to the following equations:

$$\text{Yield (\%)} = \frac{\text{Mass of starch obtained}}{\text{Initial weight of the sample}} \times 100$$

Obtaining the yield based on the initial sample content of starch:

$$\text{Yield (\%)} = \frac{\text{Mass of starch obtained}}{\text{Total mass of starch present in the sample}} \times 100$$

* The content of starch from the sample needs to be determined as described in Chapter 9 or using enzymatic kits.

After obtaining the yield of the three replicates per sample, calculate the mean, standard deviation (SD), and coefficient of variation (CV) according to equations provided by data tabulation programs, such as Microsoft Excel or Google Sheets.

The standard deviation is considered optimal when its value is as close to zero as possible. A coefficient of variation below 10% indicates a dispersion close to ideal for this type of analysis.

5 Notes

1. Fill all capped tubes with the same sample level. Furthermore, the tubes should be placed two at a time (always in pairs) on opposite sides of each other to prevent the centrifuge from becoming unbalanced and potentially stalling.
2. Resuspend the present material until no material remains at the bottom of the tube. This improves starch purification.
3. Ensure the Büchner funnel is properly connected to the filtration flask (Kitassato) and that the latter is connected to a vacuum pump or vacuum source (water horn). Avoid filling the Büchner funnel with a large amount of liquid at once, as this can overload the filter paper and cause leaks. Wash the solid retained on the filter paper with the same solvent used in the solution to remove any impurities. After filtration, remove the filter paper with the solid and dry it to obtain the final product. Thoroughly clean the funnel and filtration flask after use.
4. Avoid using high temperatures to avoid damaging the native starch., Temperatures between 30°C and 45°C are recommended.

Acknowledgments

The authors would like to thank the National Council for Scientific and Technological Development (CNPq), Brazil, for the scholarships awarded to MTP Clerici (Processes 312660/2023-5). Additionally, the authors express gratitude to the Coordination for the Improvement of Higher Education Personnel (CAPES, financial code 001), Brazil, for the scholarships awarded to NA Brandão (Process 88887.605710/2021-00), HJMC (Process 88887.141456/2025-00) and for supporting the Graduate Program in Food Technology at the Universidade Estadual de Campinas (UNICAMP), São Paulo, Brazil, and the São Paulo Research Foundation (FAPESP), Brazil (Process Number 2023/12726-8) for the postdoctoral fellowship to BL Tagliapietra.

Supplementary

Supplementary Material 1

PPM
ppm or mg/L
where:

$$200\ ppm = 200\ mg\ of\ active\ chlorine\ per\ liter\ of\ solution$$

Considering commercial sodium hypochlorite with approximately 12% active chlorine (Check the manufacturer's packaging as the active chlorine concentration can vary between 11% and 15%):

$$12 = 120{,}000\ \text{ppm}$$

Therefore

$$\text{If } 1200\ \text{ppm} \rightarrow 1\ \text{L of hypochlorite}$$

$$200\ \text{ppm} \rightarrow x$$

Therefore

$$x = 1.67\ \text{mL}$$

Add 1.67 mL of commercial 12% hypochlorite to each liter of water to obtain a 200 ppm solution.

Supplementary Material 2

Centrifuge Calibration and Operation
Detailed Steps for Calibration and Use

1. Inspection and Preparation
   - Ensure that the rotor and tubes are clean and in good condition.
   - Verify that the tubes are compatible with the rotor and free from any damage.

- Replace any tube that is cracked, damaged, or weakened.

2. Tube Balancing
   - If using microtubes, fill a balancing tube with water to match the sample mass.
   - Place sample tubes in opposite positions within the rotor to maintain balance.
   - If using multiple tubes, arrange them symmetrically to ensure even distribution.
   - Ideally, tube masses should be identical, with a maximum variance of 0.1 g.

3. Setting Parameters
   - Set the desired operational parameters, including speed, time, and temperature (if applicable).
   - To calibrate the speed, use a tachometer or optical sensor to measure actual versus set speed.
   - Calibrate the timer by comparing the real elapsed time with the programmed duration.
   - If the centrifuge has temperature control, verify the actual temperature using a calibrated thermometer and adjust as needed.

4. Operation and Maintenance
   - Ensure that all tubes are securely fitted into the adaptors with no looseness.
   - Avoid sudden movements when inserting or removing tubes to prevent mixing or imbalance.
   - If the centrifuge lacks an automatic safety lock, wait until the rotor comes to a complete stop before opening the lid.
   - Keep the centrifuge clean and free of debris, following the manufacturer's maintenance instructions.

Importance of Proper Balancing: Proper balancing of the tubes is critical to prevent excessive vibration and to ensure the safety of both the equipment and laboratory personnel. Unbalanced tubes may damage the rotor and pose a risk of accidents.

Additional Resources: For more information, consult the user manual: https://kasvi.com.br/uso-centrifuga-laboratorio/

Supplementary Material 3

Table 3
Correspondence between the diameters of the filter papers and the funnels (60°) that ensures filtration under the best conditions

Funnel diameter (maximum opening) (mm)	Filter paper diameter (cm)
35	5.5
45	7.0
55	9.0
65	11.0
75	12.5
90	15.0
160	24.0

Reference

1. American Association of Cereal Chemists (2010) AACCI approved methods of analysis, 11th edn. AACC International, St. Paul

Chapter 7

Starch Extraction by the Reducing Method

Hugo José Martins Carvalho and Maria Teresa Pedrosa Silva Clerici

Abstract

Laboratory extraction methods are essential for obtaining native starch in adequate quantity and quality for characterization, especially from underexplored raw materials. Aqueous extraction techniques and those with a reducing agent, such as active SO_2, are used when raw materials have a high protein content and/or enzymatic browning enzymes, making starch separation difficult and promoting the formation of brownish starches, respectively. Some raw materials also contain carbohydrates and proteins that can undergo fermentation during maceration in water, so bisulfite inhibits microorganisms. This chapter covers the main methods, as well as precautions, standardization, and strategies to optimize the yield and purity of the extracted starch.

Keywords Isolation, Native starch, Purity

1 Introduction

Starch extraction involves the separation of starch granules from plant tissues and is a critical step to ensure high-purity starch while preserving its native properties. The primary industrial and laboratory sources of starch are cereal seeds, such as maize, wheat, and rice, and underground storage organs (roots, tubers, and rhizomes), found in materials like potato, cassava, and taro. While cereals typically exhibit low moisture content (10%–14%) and a high starch concentration in the endosperm, underground organs contain higher moisture levels (≥60%) and lower lipid and protein contents.

The choice of extraction method can directly impact parameters such as starch purity, yield, as well as structural aspects like granule integrity and technological or nutritional properties. Major challenges include the presence of nonstarch components such as mucilages, phenolic compounds, proteins, and fibers, which may hinder both extraction efficiency and the functional characteristics of the final starch product. Therefore, selecting an appropriate extraction method and fine-tuning parameters such as mechanical force, processing time, and the use of agents (reducing agents) are essential

Maria Teresa Pedrosa Silva Clerici and Pedro H. Campelo (eds.), *Basic Protocols in Starch Characterization*, Methods and Protocols in Food Science, https://doi.org/10.1007/978-1-0716-5116-2_7,

for process optimization, whether at laboratory or industrial scale, to obtain starches with properties suitable for food and industrial applications.

The most commonly used starch extraction methods across various matrices are aqueous extraction, which relies on water and mechanical processing, and extraction using reducing agents such as SO_2. These agents promote more effective removal of nonstarch polysaccharides, proteins, and associated compounds, resulting in a higher degree of starch purity.

2 Material

1. Disposable gloves.
2. Glass beakers (50 mL, 600 mL, and 1 L).
3. Metal spatula.
4. Filter mesh (88 mesh).
5. 50 mL Falcon tubes with caps, previously weighed and labeled (tube and cap).
6. Distilled water wash bottle.
7. Petri dish.
8. Mortar and pestle.

2.1 Reagents

1. Freshly distilled deionized water.
2. Sodium hypochlorite (NaClO) is used exclusively for sample sanitization.
3. Reducing agent—sodium metabisulfite ($Na_2S_2O_5$) or sodium bisulfite ($NaHSO_3$).
4. pH 4.0 and pH 7.0 buffer solutions for pH meter adjustment.
5. 0.01 N HCl.

3 Method

Starch extraction using a reducing agent [1].

3.1 Sample

Approximately 15 g of fresh sample (grains, seeds, fruits, tubers, roots, and rhizomes).

3.2 Equipment

1. Analytical balance with precision up to 0.0001 g, within calibration period and with available standard operating procedure.
2. Drying oven with air circulation.

3. Rotating centrifuge for 50 mL Falcon-type tubes with a cooling system.
4. Mixer or blender, with an available standard operating procedure.
5. Water bath (30–60°C).
6. pH meter, calibrated at pH 4.0 and pH 7.0.
7. Mixer or blender.

3.3 Procedure

3.3.1 Preanalysis Steps

1. Preselect the number of beakers according to the number of samples and perform the analyses in triplicate. Check that they are prelabeled and resistant to washing and heating. If not, use a high-temperature pen to identify them.
2. Check the operating procedure of the air circulation oven and whether it is precalibrated. Record this information. Strictly follow the operating procedure.
3. Check the standard operating procedure of the analytical balance and whether it is calibrated. Record the calibration date. Strictly follow the operating procedure.
4. Check the standard operating procedure of the pH meter and whether it is calibrated. Strictly follow the operating procedure.
5. To collect data, prepare a results table in the laboratory notebook, as shown in Table 1 in the **Results**.

3.3.2 Starch Extraction

1. Begin by washing the sample under running water to remove dirt, stones, branches, or other debris (applicable to grains, seeds, fruits with peel, *Solanum lycocarpum*, banana, tubers, roots, and rhizomes).
2. Sanitize the samples using a 200 ppm sodium hypochlorite solution. The preparation for this concentration is described in Supplementary Material 1 in Chapter 6.
3. Immerse the samples in the sanitizing solution and allow them to rest for 15 min. Ensure that the sanitizer fully covers the samples and gently stir every 5 min during this period.
4. After the sanitization time, discard the sanitizing solution.
5. Rinse the samples thoroughly using freshly distilled deionized water to remove residual chlorine.
6. Repeat the rinsing process at least three times to ensure adequate chlorine removal.

7. If necessary, allow the samples to air dry overnight in a sieve or draining rack to remove remaining hypochlorite through evaporation.
8. Weigh the samples using an analytical balance (precision 0.01 g), using approximately 10–15 g, and record the exact weight.*
   **The entire process should be performed in triplicate.*
9. If required, peel the samples using a knife.
10. Transfer the weighed samples to a 600 mL beaker.
11. Add freshly distilled water in a 1:3 ratio (sample:water) for processing.

    Example:
    For every 100 g of sample to be used:

    100 g of sample → ratio of 1

    Thus

    1 → 100 g of sample
    3 → x g of water

    Therefore

    $x = 300$ g of distilled water◀
12. Add 0.2% (sample basis) of SO_2 to the beaker containing the grain and distilled water.
13. The amount of SO_2 in grams to be added should follow the calculations shown in Supplemental Material 1.
14. Place the beakers in a previously calibrated pH meter.
15. Correct the pH by adding 0.01 N HCl solution, adjusting it to a pH range between 3 and 4.
16. After correcting the pH, cover the beakers with PVC plastic wrap (to prevent dirt, insects, and other materials from falling in during the extraction process).
17. Use a spatula to make holes in the previously applied coating to facilitate the release of hot air.
18. Place in a water bath preheated to 50°C.
19. Keep in the water bath for 24–48 h (*see* **Note 1**)
20. After the maceration period under heating, remove the sample from the water bath, as illustrated in Fig. 1.
21. Remove the liquid using a sieve.
22. Wash with a pestle containing freshly distilled water to remove any residual SO_2.
23. Repeat the previous step at least three times to properly remove any residual reducing agent.
    Just for grains like corn:

Fig. 1 (**A**) Unmacerated fresh corn containing SO_2 and (**B**) corn macerated at 24 h/50°C. Source: Authors' collection, 2025

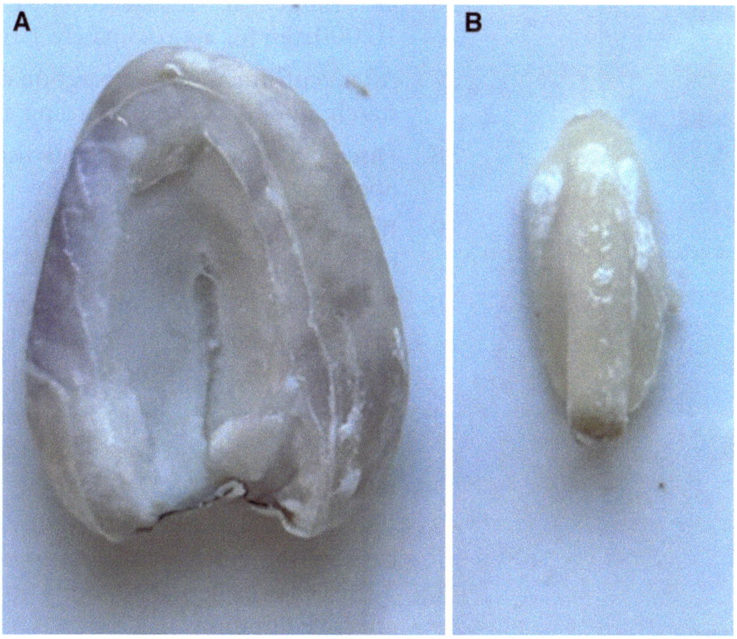

Fig. 2 Macerated corn grain (**A**) and germ (**B**)

24. Transfer the grains to the mortar and pestle.
25. Press the pestle onto the grains.
26. Using your hands (wearing gloves), separate the germ from the grain, as shown in Fig. 2 (*see* **Note 2**).

27. After separating the fractions reported in the previous item, dispose of the germ in an appropriate place.
    Resuming application in all samples to have starch extracted
28. The sample should be taken to a processor/mixer.
29. Add freshly distilled water in a ratio of 1:5.
30. Process for 1 to 2 min at medium speed.
31. Transfer the mixture to a filtration mesh (88 mesh) for sieving.
32. Rinse the processing container using a wash bottle with small amounts of distilled water to recover any retained material, and proceed with filtration.
33. Transfer the filtrate to 50 mL Falcon tubes and seal with the corresponding caps.
34. Transport the tubes containing the starch to the centrifuge.*
    **Refer to Supplementary Material 2 in Chapter 6 for detailed calibration and operational instructions for the centrifuge.*
35. Place tubes with similar weights in opposite positions in the rotor to ensure the center of gravity remains balanced.
36. Centrifuge at refrigeration temperature (~4–6°C) at 10,000 rpm for approximately 10 min.
37. After centrifugation, a separation of layers will be visible, with starch, due to its density settling at the bottom of the tube.
38. Discard the supernatant, and using a spatula, remove the upper layer (containing proteins/fibers/lipids).
39. Resuspend the sediment at the bottom of the tube in distilled water (*see* **Note 3**).
40. Repeat steps 37–39 until only starch remains in the tube and the supernatant water appears clear, free of coloration and suspended materials such as small fibers.
41. Resuspend the material in distilled water again and proceed to filtration using a Büchner funnel system connected to a vacuum pump.
42. Use a filter paper with a slightly smaller diameter than the base of the Büchner funnel to ensure a proper fit. High-quality filter papers are essential to prevent tearing and ensure effective filtration.
43. Place a preweighed filter paper (previously dried in an oven and labeled in pencil) into the Büchner funnel, ensuring it is evenly seated on the porous surface of the funnel.

44. Add a small amount of the solvent or solution to be filtered onto the filter paper. This helps secure the paper to the funnel, sealing the edges and preventing liquid and solids from passing underneath.
45. Begin applying vacuum to the filtration flask (Kitasato) connected to the Büchner funnel. The vacuum will draw the liquid through the filter paper, retaining the solids on the paper's surface (*see* **Note 4**).
46. Place the filter paper with the sample in a Petri dish and place it in the oven.
47. Dry in an air circulation oven at approximately 30–45°C with open air circulation to remove residual water overnight (~12 h) (*see* **Note 5**).
48. Place the plates in a desiccator containing silica gel (see moisture chapter) and wait 40 min before weighing.
49. Weigh the starch obtained on an analytical balance.
50. Record all values on the data recording spreadsheet (Table 1).

3.4 Results

The yield of extracted starch can be expressed by considering the amount of grain used initially, as well as the initial starch content contained in the raw material.

Calculate the total starch yield obtained according to the following equations:

$$\text{Yield (\%)} = \frac{\text{Mass of starch obtained}}{\text{Initial weight of the sample}} \times 100$$

Obtaining the yield based on the initial sample content of starch:

$$\text{Yield (\%)} = \frac{\text{Mass of starch obtained}}{\text{Total mass of starch present in the sample}} \times 100$$

After obtaining the yield of the three replicates per sample, calculate the mean, standard deviation (SD), and coefficient of variation (CV) according to equations provided by data tabulation programs, such as Microsoft Excel or Google Sheets.

The standard deviation is considered optimal when its value is as close to zero as possible. A coefficient of variation below 10% indicates a dispersion close to ideal for this type of analysis.

Table 1
Spreadsheet recording data collected during the analysis

Starch yield determination based on initial sample (%)										
Sample										
Analyst										
Date										
Sample identification (code and reference)	Number	Replica	Empty filter paper weight (g) (P1)	Initial sample weight (g) (P2)	1st Weighing crucible weight + dry starch (g) (P3)	Final starch weight (g) P3 - P1	Yield (%)	Average (%)	D.P.	CV
CS*	1	CS1								
	2	CS2								
	3	CS3								

*Example of data recording for a starch sample. *CS*, cassava starch; *SD*, standard deviation; *CV*, coefficient of variation

4 Notes

1. Check the water level before turning on and ensure that the amount of water is sufficient to cover the volume occupied in the beaker.
2. If the film on the grains is easy to remove, it can also be removed, making purification easier in later stages.
3. Resuspend the present material until no material remains at the bottom of the tube. This improves starch purification.
4. Ensure the Büchner funnel is properly connected to the filtration flask (Kitassato) and that the latter is connected to a vacuum pump or vacuum source (water horn). Avoid filling the Büchner funnel with a large amount of liquid at once, as this can overload the filter paper and cause leaks. Wash the solid retained on the filter paper with the same solvent used in the solution to remove any impurities. After filtration, remove the filter paper with the solid and dry it to obtain the final product. Thoroughly clean the funnel and filtration flask after use.
5. Avoid using high temperatures to avoid damaging the native-native starch. Temperatures between 30°C and 45°C are recommended.

Acknowledgments

The authors would like to thank the National Council for Scientific and Technological Development (CNPq), Brazil, for the scholarships awarded to MTP Clerici (Processes 312660/2023-5). Additionally, the authors express gratitude to the Coordination for the Improvement of Higher Education Personnel (CAPES, financial code 001), Brazil, for the scholarships awarded to HJMC (Process 88887.141456/2025-00) and for supporting the Graduate Program in Food Technology at the Universidade Estadual de Campinas (UNICAMP), São Paulo, Brazil.

Supplementary Material

Supplementary Material 1

Active SO_2 calculation

Determination of the amount of SO_2 to be added for starch extraction:

Considering that the reagent to be used is sodium metabisulfite ($Na_2S_2O_5$):

Molar mass of $Na_2S_2O_5$:

Na : 22.99 → considering that there are 2 sodium molecules present: 22.99 × 2 = 45.88 g/mol

S : 36.06 → considering that there are 2 sulfur molecules present: 32.06 × 2 = 64.12 g/mol

O : 16 → considering that there are 5 oxygen molecules present: 16 × 5 = 80 g/mol

Therefore, totaling 190.11 g/mol per $Na_2S_2O_5$ molecule and considering that the molar mass of SO_2 is:

$$32.06 + 2 \times 16 = 64.06 \text{ g/mol}$$

Each mole of $Na_2S_2O_5$ releases 2 moles of SO_2, or:

$$190.11\text{g/mol de } Na_2S_2O_5 \rightarrow 2 \times 64.6 \text{ g/mol de } SO_2 = 128.12 \text{ g/mol of } SO_2$$

present in sodium metabisulfite

Therefore, the conversion factor can be given by:

$$SO_2 = \frac{\text{Molar mass of } SO_2}{\text{Molar mass of } Na_2S_2O_5} = \frac{128.12}{190.10} = \sim 0.674$$

or 67.4% of the $Na_2S_2O_5$ molecule is active SO_2

Considering a purity of 97% of the product:

$$SO_2 = 0.97 \times 0.674 = 0.654 \text{ or } 65.4\% \text{ of active } SO_2$$

Considering that 0.2% of active SO_2 will be added to 10 g of sample:

100 g → 65.4%

x → 0.2%

X = 0.3058 g of $Na_2S_2O_5$ for every 100 g of sample

Therefore, for 10 g:

100 g → 0.3058

10 g → x

X = 0.0305 g to achieve 0.2% action in 10 g of sample.

Reference

1. American Association of Cereal Chemists (2010) AACCI approved methods of analysis, 11th edn. AACC International, St. Paul

Check for updates

Chapter 8

Procedure for Color Analysis of Different Starches Using a Portable Colorimeter

Josias Martins dos Anjos Cruz, Edgar Aparecido Sanches, Pedro Henrique Campelo and Jaqueline de Araújo Bezerra

Abstract

Color is one of the most important evaluation parameters, as starch is considered an opaque white powder, so if it has changes in color, this fact must be investigated regarding its purity and extraction method. Standardizing color analysis using a colorimeter must be performed to ensure a reliable starch quality result.

Keywords Starch, Color, Analytical method

1 Introduction

Starch is a natural, biodegradable, and inexpensive polysaccharide, being the second most abundant organic substance in nature, behind only cellulose. Sources of starch include cereals, legumes, tubers, and green fruits [1]. This biopolymer mainly consists of amylose and amylopectin, and its granular morphology depends on the source from which it is extracted [2].

Digital color analysis has been widely used in different fields of knowledge, making it necessary to develop fast and reliable methods [3]. Currently, various devices are used for color detection, such as sensors [4], digital analyzers [5], and portable colorimeters [6, 7].

A wide variety of color spaces are currently used for color quantification, such as RGB, HSV, CIE XYZ, and CIE Lab [8]. The CIE Lab color space has been widely used in various industries to represent color results [9–11].

The CIE Lab color space has three variables that correspond to the L^* parameter, representing lightness (positive values indicate higher brightness and negative values indicate darker), and a^* and

Maria Teresa Pedrosa Silva Clerici and Pedro H. Campelo (eds.), *Basic Protocols in Starch Characterization*, Methods and Protocols in Food Science, https://doi.org/10.1007/978-1-0716-5116-2_8,

b^* that describe the hue of the color from green (−) to red (+), and from blue (−) to yellow (+), respectively [9]. Obtaining the values for each parameter can easily plot a graph in rectangular coordinates.

Alternatively, the CIE LCh color scale can also be used, equivalent to the CIE Lab space but with better adjustment to color perception by human eyes [12]. This color scale uses cylindrical coordinates; the L^* parameter indicates lightness, C^* represents chroma or saturation, indicating the distance from the lightness axis, and finally, h represents the hue angle, expressed in degrees from + a counterclockwise [13].

This study aims to establish a procedure for color analysis of starches using a portable colorimeter.

2 Material

For color analysis, two starch samples were selected: one conventional and one unconventional. The conventional sample, corn starch, was purchased from the market. The unconventional starch sample, ariá starch, underwent processing and was then dried and sieved for analysis.

3 Method

3.1 Experimental Procedure

The analyses were performed using the Delta Vista portable colorimeter (Delta Color, Brazil), duly calibrated. Figure 1 outlines the calibration procedure for this equipment. Whenever there is a change in the sample holder opening, this procedure must be performed. The desired opening must be selected in step 3 according to Fig. 1. In this work, we used openings of 4 and 6 mm.

After calibration, the equipment was connected via USB to a notebook with installed software (i7 Gold Software, version 1.0.5.5 ×64), and sample information was added, as shown in Fig. 2. The samples were in the form of fine, dry powder. A cut tissue paper (10 cm × 10 × 10 cm) was placed on a flat surface, and a portion of the sample sufficient to cover the surface of the paper and the 4 or 6 mm area of the colorimeter sample holder was added. The equipment was then placed over the sample holder (Fig. 3).

The analyses were acquired in triplicate with the 4 mm sample holder opening (Fig. 4). After that, the entire procedure was repeated from calibration for the 6 mm opening. The color parameter results L^*, a^*, b^*, C^*, and h^* were expressed as mean ± standard deviation.

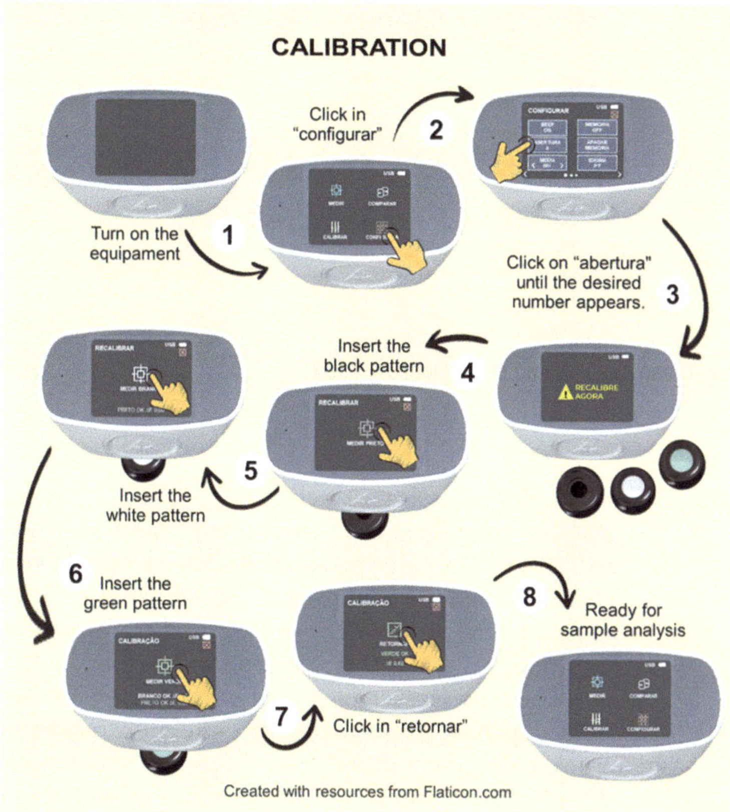

Fig. 1 Schematic for the calibration of the Delta Vista Portable colorimeter. Source: Authors' collection, 2025

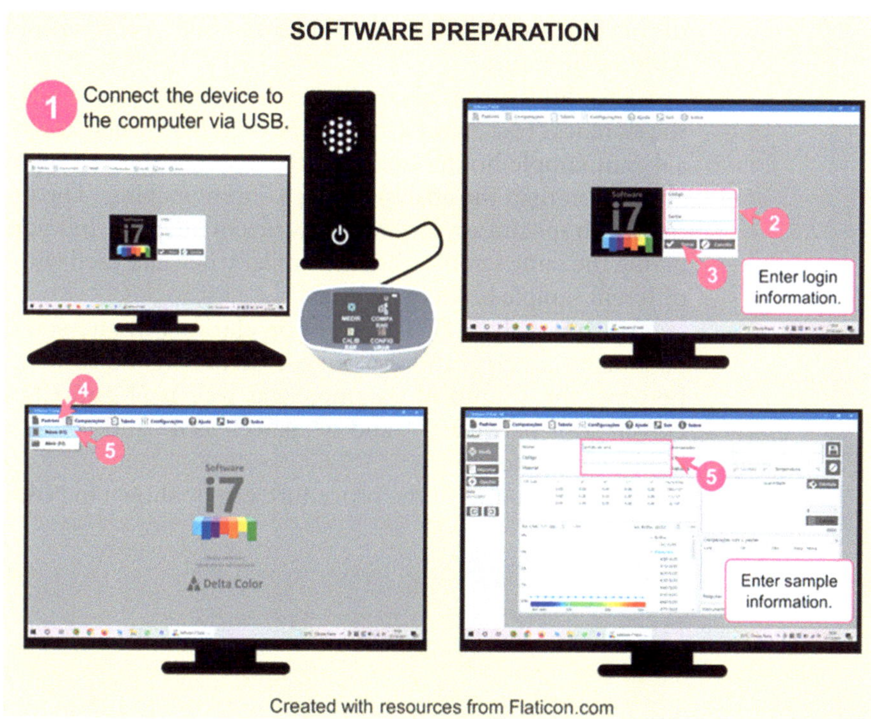

Fig. 2 Connection of the colorimeter and setup of the i7 Gold Software. Source: Authors' collection, 2025

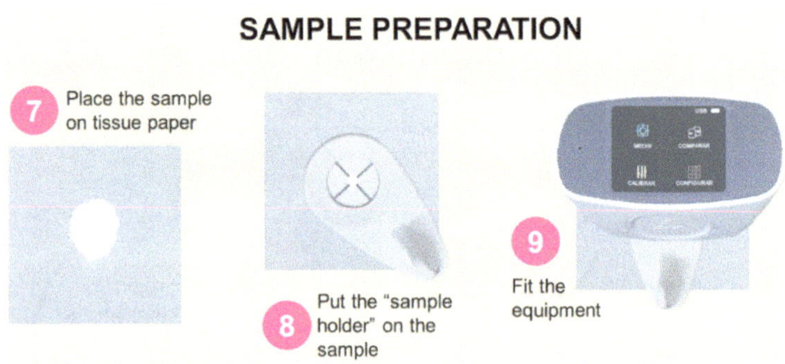

Fig. 3 Sample preparation. Source: Authors' collection, 2025

3.2 Results and Discussion

The color analysis parameters L^*, a^*, b^*, C^*, and h^* for the two starches using the 4 and 6 mm sample holder openings are described in Table 1. The color parameters showed high L^* values, indicating that the samples are quite light. Additionally, a^*, b^*, and C^* values close to zero indicate the color trend toward the central axis of lightness, resulting in the perception of white.

The analyses were acquired in triplicate with the 4 mm sample holder opening (Fig. 4). After that, the entire procedure was repeated from calibration for the 6 mm opening. The color parameter results L^*, a^*, b^*, C^*, and h^* were expressed as mean ± standard deviation.

The results indicate that, except for the C^* parameter of ariá, the starches analyzed with different sample holder openings have statistically different parameters. Therefore, it is necessary to report the opening used in the methodology to enable a correct comparison with the literature.

This difference is also evidenced by the spectrum and color of the starches (Fig. 5), where it can be seen that, when analyzed with a 4 mm sample holder opening, the starches showed higher color absorption than the analysis using a 6 mm opening. The profiles of the starch spectra are different in terms of intensity, but when comparing the same sample, they are similar when analyzed with the two different sample holder openings.

In the analysis of Pearson correlations of the starch color parameters for corn and ariá, there is an excellent direct correlation between the L^* and h^* parameters as well as b^* and C^*; indirect correlations between L^* and a^* as well as a^* and h^* as shown in Table 2 in **Notes**.

The conclusion of the procedure can be shown in **Note 1**.

Table 1
Starch color

Starch	Sample holder	Color parameters				
		*L**	*a**	*b**	*C**	*h**
Corn	4 mm	87.0 ± 0.5[c]	−0.1 ± 0.0[b]	3.5 ± 0.1[b]	3.5 ± 0.1[b]	91.4 ± 1.0[b]
	6 mm	93.3 ± 0.4[a]	−2.0 ± 0.2[d]	4.5 ± 0.0[a]	4.9 ± 0.1[a]	112.4 ± 3.0[a]
"Ariá"	4 mm	86.0 ± 0.4[c]	0.2 ± 0.0[a]	2.2 ± 0.1[c]	2.2 ± 0.1[c]	84.7 ± 1.5[c]
	6 mm	91.4 ± 0.4[b]	−0.9 ± 0.0[c]	1.9 ± 0.1[d]	2.1 ± 0.1[c]	116.2 ± 1.8[a]

±Means that do not share the same letter are significantly different

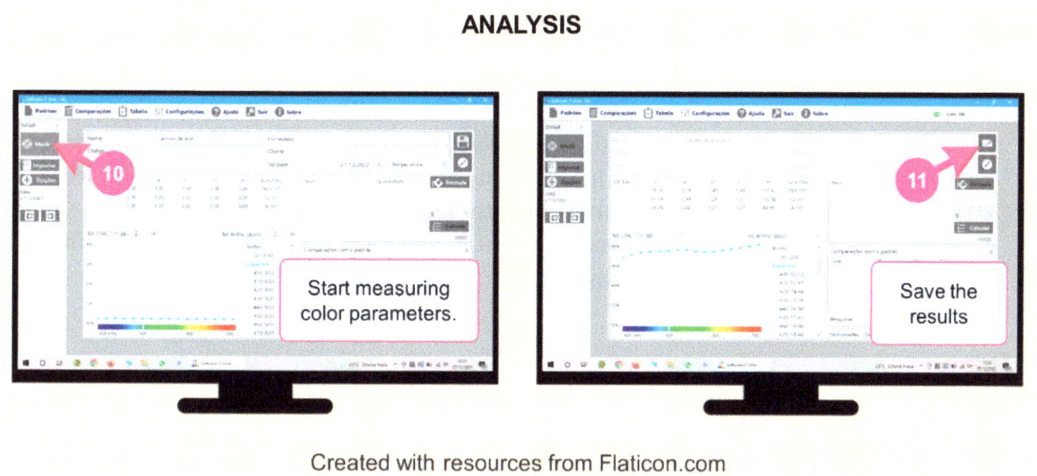

Fig. 4 Analysis of the samples in the i7 Gold Software. Source: Authors' collection, 2025

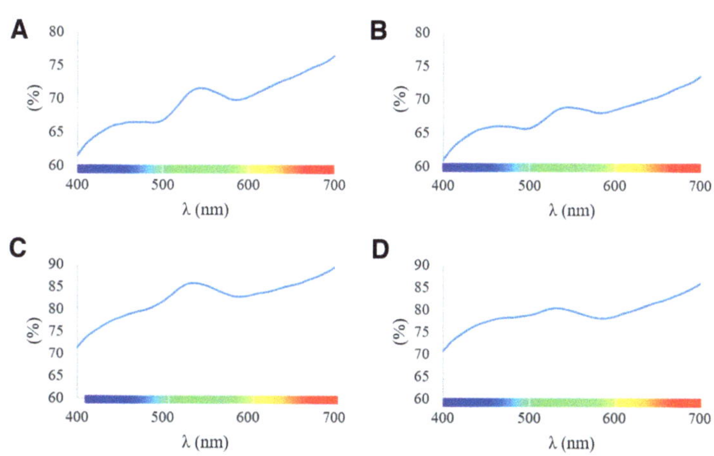

Fig. 5 Color spectrum of the starches: (**A**) corn 4, (**B**) corn 6, (**C**) ariá 4, and (**D**) ariá 6. Source: Authors' collection, 2025

Table 2
Pearson correlations

	*	*a**	*b**	*C**
*a**	−0.959			
*b**	0.422	−0.613		
*C**	0.527	−0.708	0.992	
*h**	0.935	−0.844	0.170	0.280

4 Note

1. The presented color parameter data for corn and ariá starches are novel, providing a primary source for comparison and discussion of results. The procedure presented here is a viable and easy methodology for determining the color parameters of starches from different matrices. The main observation is the statistical difference between the starch color parameters when using different sample holder openings. For future studies, it is important to indicate the opening used in the sample analysis and, if possible, perform it with different openings for comparison.

References

1. Ashogbon AO, Akintayo ET (2014) Recent trend in the physical and chemical modification of starches from different botanical sources: a review. Starch 66(1–2):41–57
2. Pérez S, Bertoft E (2010) The molecular structures of starch components and their contribution to the architecture of starch granules: a comprehensive review. Starch 62(8):389–420. https://doi.org/10.1002/star.201000013
3. Tiuftiakov NY, Kalinichev AV, Pokhvishcheva NV, Peshkova MA (2021) Digital color analysis for colorimetric signal processing: towards an analytically justified choice of acquisition technique and color space. Sens Actuators B 344:130274
4. Gavrilenko NA, Muravyov SV, Silushkin SV, Spiridonova AS (2014) Polymethacrylate optodes: a potential for chemical digital color analysis. Measurement 51(1):464–469
5. Hirayama E, Sugiyama T, Hisamoto H, Suzuki K (2000) Visual and colorimetric lithium ion sensing based on digital color analysis. Anal Chem 72(3):465–474
6. Dang DS, Buhler JF, Stafford CD, Taylor MJ, Shippen JE, Dai X, England EM, Matarneh SK (2021) Nix Pro 2 and Color Muse as potential colorimeters for evaluating color in foods. LWT Food Sci Technol 147:111648
7. Milovanovic B, Tomovic V, Djekic I, Miocinovic J, Solowiej BG, Lorenzo JM, Barba FJ, Tomasevic I (2021) Colour assessment of milk and milk products using computer vision system and colorimeter. Int Dairy J 120:105084
8. Capitán-Vallvey LF, López-Ruiz N, Martínez-Olmos A, Erenas MM, Palma AJ (2015) Recent developments in computer vision-based analytical chemistry: a tutorial review. Anal Chim Acta 899:23–56
9. Bell SC, Nawrocki HD, Morris KB (2009) Forensic discrimination of glass using cathodoluminescence and CIE LAB color coordinates: a feasibility study. Forensic Sci Int 189(1–3):93–99
10. Kulapichitr F, Borompichaichartkul C, Fang M, Suppavorasatit I, Cadwallader KR (2022) Effect of post-harvest drying process on chlorogenic acids, antioxidant activities and CIE-Lab color of Thai Arabica green coffee beans. Food Chem 366:130504

11. Vankar PS, Shukla D (2019) Dyeing application of newer natural dyes on cotton silk and wool with fastness properties, CIE lab values, and shade card. In: New trends in natural dyes for textiles. Elsevier, pp 159–282
12. Xi W, Liu Y, Zhao W, Hu R, Luo X (2021) Colored radiative cooling: how to balance color display and radiative cooling performance. Int J Therm Sci 170:107172
13. Konica Minolta Sensing Americas (2020) Understanding the CIE L*C*h Color Space. https://sensing.konicaminolta.us/us/blog/understanding-the-cie-lch-color-space/ Accessed 9 Apr 2025

Check for updates

Chapter 9

Detection and Characterization of Starch by Near Infrared Spectroscopy

Amanda Teixeira Badaró, Luis Jam Pier Cruz-Tirado, Jose Blasco, Nuria Aleixos-Borrás and Douglas Fernandes Barbin

Abstract

Starch is a very important and common polysaccharide found in food and organic products. The most common standard methods to determine starch content in food are those established by the Association of Official Analytical Chemists (AOAC) or the American Association of Cereal Chemists (AACC). The determination of starch in foodstuffs is time-consuming and uses a lot of reagents, so the development of techniques that are faster and do not require sample preparation is interesting. Recently, distinct instrumental methods have been used in food analyses, as near infrared (NIR) spectroscopy and hyperspectral imaging (HSI). However, working with spectroscopy can be a challenge. Thus, this chapter shows a tutorial on NIR spectroscopy and hyperspectral imaging data acquisition for the detection and characterization of starch.

Keywords NIR, HSI, Spectroscopy

1 Introduction

Starch is a very important and common polysaccharide found in food and organic products. Naturally, its form is water-insoluble, but its structure can be modified in order to achieve desirable characteristics. It is known that starches contribute a lot in food processing, and they are widely used as food additives, for example, to modify the texture of gels. Therefore, it is important to analyze both the content and the type of starch present in the food. Another important aspect of starches is their content of amylose and amylopectin, which can have a great influence on the final product [1, 2].

The most common standard methods to determine starch content in food are those established by the Association of Official

Maria Teresa Pedrosa Silva Clerici and Pedro H. Campelo (eds.), *Basic Protocols in Starch Characterization*, Methods and Protocols in Food Science, https://doi.org/10.1007/978-1-0716-5116-2_9,

Analytical Chemists (AOAC) or the American Association of Cereal Chemists (AACC) [3, 4]. Both institutions have procedures adopted for the different types of starchy foods. However, an essential step is needed before the determination of starch content: it is necessary to separate the starch from other constituents present in the food matrix, since its semicrystalline structure does not allow the access of reagents [1].

Therefore, it is clear that the determination of starch in foodstuffs is time-consuming and uses a lot of reagents. Apart from that, in some quantitative analyses , the starch extracted and quantified as the amount of starch present in the food matrix is determined as glucose, so the presence of other sugars can affect the final result. This shows the need for trained personnel to carry out the analyses [1].

Considering all these drawbacks, the development of techniques that are faster and do not require sample preparation is interesting. Recently, distinct instrumental methods have been used in food analyses. Among them, we will mention the spectral techniques, more specifically, near infrared (NIR) spectroscopy and hyperspectral imaging (HSI).

NIR spectroscopy is a vibrational technique used to determine the chemical composition of organic samples. When a sample is exposed to radiation in the NIR range, there are overtones and combinations of chemical bonds of C–H, N–H, O–H, and S–H which absorb the energy in specific wavelengths. NIR spectra are taken from single points that may not be representative of the whole sample if it is a heterogeneous matrix [5]. In this context, there is a suitable alternative that has been increasingly used: hyperspectral imaging (HSI). HSI combines spatial and spectral information of a sample, providing a spectrum to each pixel of an image, and, consequently, detailed information of a component on the sample surface [6].

NIR spectroscopy and hyperspectral imaging are able to provide relevant and reliable information about food quality, process control, and so on. Usually, it is necessary to combine the traditional methods of analysis and the spectral techniques in order to develop the calibration models. However, once the models are created, it is easy to use, and there is no need for long sample preparation, providing the identification of different starch parameters from the same dataset simultaneously.

The choice of which method is best depends on many factors. Prior to any analysis, it is important to know your sample and the parameters you want to determine. It is necessary to understand the object of study in order to select the right tools (i.e., type of equipment, spectral range, and acquisition mode) to obtain reliable results. Working with spectroscopy can be a challenge. Thus, this chapter shows a tutorial of NIR spectroscopy and hyperspectral imaging (HSI) data acquisition for the detection and characterization of starch.

2 Materials

1. If samples are in powder form, sieve the samples to standardize sample particle size (*see* **Note 1**).
2. If samples are hard (rigid or crusty), analysis can be carried out using the whole sample or samples can be ground to a powder form.

3 Methods

3.1 Sample Preparation

1. Vigorously mix the samples prior to any analysis (*see* **Note 2**).
2. It is suggested to place samples in Petri dishes or vials with no absorbance/reflectance in the NIR region.
3. Make sure to compress the samples if they are in powder form.
4. The number of samples is very important. However, there is no right number; the best number of samples is the one necessary to obtain reliable models. This can be determined by the figures of merit of your models (*see* Sect. 3.4.8).
5. The sample set must contain samples with all the variations found in your object of study.

3.2 Near Infrared Spectrometer

1. First, calibrate the equipment using a Spectralon® (*see* **Note 3**).
2. The equipment must be placed in a dark chamber to minimize external light.
3. A spectrum of a white sample (Spectralon®) must be acquired to correct for light effects (*see* **Note 4**).
4. If the spectrometer is portable, the equipment must be placed on the surface of the sample holder. In this case, only one spectrum is acquired for each sample; therefore, it is suggested to acquire spectra from different points of the sample. All the spectra can be used, or they can be averaged, and the mean spectrum used as a representative of the sample.
5. If the spectrometer is a bench type, the equipment usually includes a sample holder and spectra are automatically acquired at different points of the sample.

3.3 Near Infrared Hyperspectral Imaging System

1. First, calibrate the equipment using a Spectralon® (*see* **Note 3**).
2. If the hyperspectral imaging system does not automatically subtract the white and dark references from the images, white and dark images must be acquired for further image correction. The white image (>99% reflectance) can be acquired by using a Spectralon®, and the dark image (0% reflectance) can be acquired in a dark chamber with all lights off and the camera lens covered.
3. If the equipment has the automatic correction of white and dark references, step 2 is not necessary.
4. The parameters, such as the distance between the sample and the camera or the light source, must be set and maintained for all samples.
5. Adjust the exposure time, the conveyor speed and use the correct lens according to your samples and parameters to be determined.
6. In case of opaque samples, the reflectance or absorbance mode is used for data acquisition. In the case of translucent samples, transmission mode is used. In this case, the light passes through the sample, and the time of exposure may not be enough to obtain an image, so a calibration must be done. The calibration is carried out by increasing the integration time as much as possible without reaching the maximum intensity (saturation) in any wavelength in any part of the sample. Then, a calibration file is created and used for image acquisition.

3.4 Data Processing

3.4.1 Spectra and Image Correction

1. As mentioned in Sect. 3.3, step 2, if the equipment does not automatically correct the images with white and dark references, this step must be carried out to calculate a relative image. The relative image or spectrum (R) is calculated according to the following equation:

$$R = \frac{R_0 - R_D}{R_W - R_D} \tag{1}$$

where R_0 is the original hyperspectral image (or spectrum), R_D is the dark image (or spectrum), and R_W is the white reflectance image (or spectrum).

3.4.2 Image Segmentation and Spectra Extraction

1. Region of interest (ROI) extraction and image segmentation must be performed prior to data analysis. ROI are determined by thresholding the images in order to remove undesirable information as the background or the border of the sample holder.

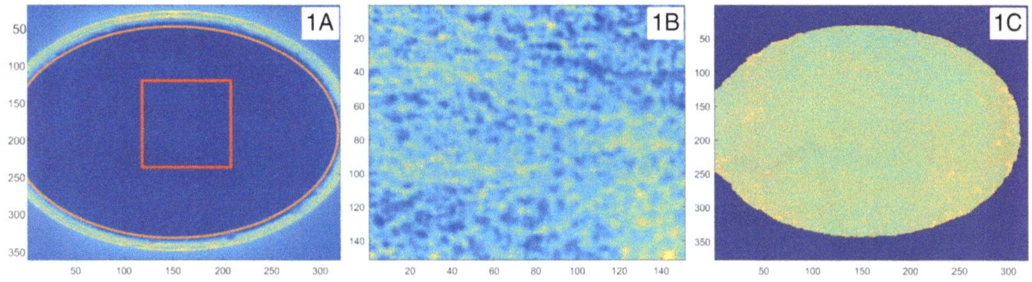

Fig. 1 Hyperspectral image (**1A**), square ROI (**1B**), and ROI removing only undesirable information (**1C**). Source: Authors' collection, 2025

2. The design of the ROI can be chosen according to your sample. It can be selected as square regions, round regions, random pixels, or other methods. The best ROI is the one that has all the information and variability of the sample. Figure 1A shows an example of an NIR image of wheat flour, and Fig. 1B, C shows two different ROI selected from this image.
3. From the ROIs, two paths can be followed: (A) extraction of a mean spectrum from the ROI or (B) working pixel by pixel.

3.4.3 Sample Selection

1. Prior to data analysis, the dataset must be split into two groups: calibration and validation sets. Normally, calibration models are developed with an internal validation method, called cross-validation. However, an external validation must be performed by using a group of samples that is not included in the calibration set.
2. Usually, the dataset is split into 75% for calibration (or a range between 60% and 80%, depending on the number of samples available) and 25% for validation, as an external set of samples (it may be in a range between 40% and 20%, according to the number of samples available).

3.4.4 Spectra and Image Preprocessing

1. Spectra/image preprocessing must be performed in order to reduce the influence of nonrelated information on the target analysis. For instance, there is the influence of light source, particle size, etc., and some methods help to reduce this influence.
2. Multiplicative scatter correction (MSC), standard normal variate (SNV), and first and second order derivatives are commonly used in this sense. Figure 2 shows an example of preprocessing techniques applied to flour's spectral information.
3. Most of the statistical software used for multivariate statistical analysis has these tools and can be easily applied to the spectral data.

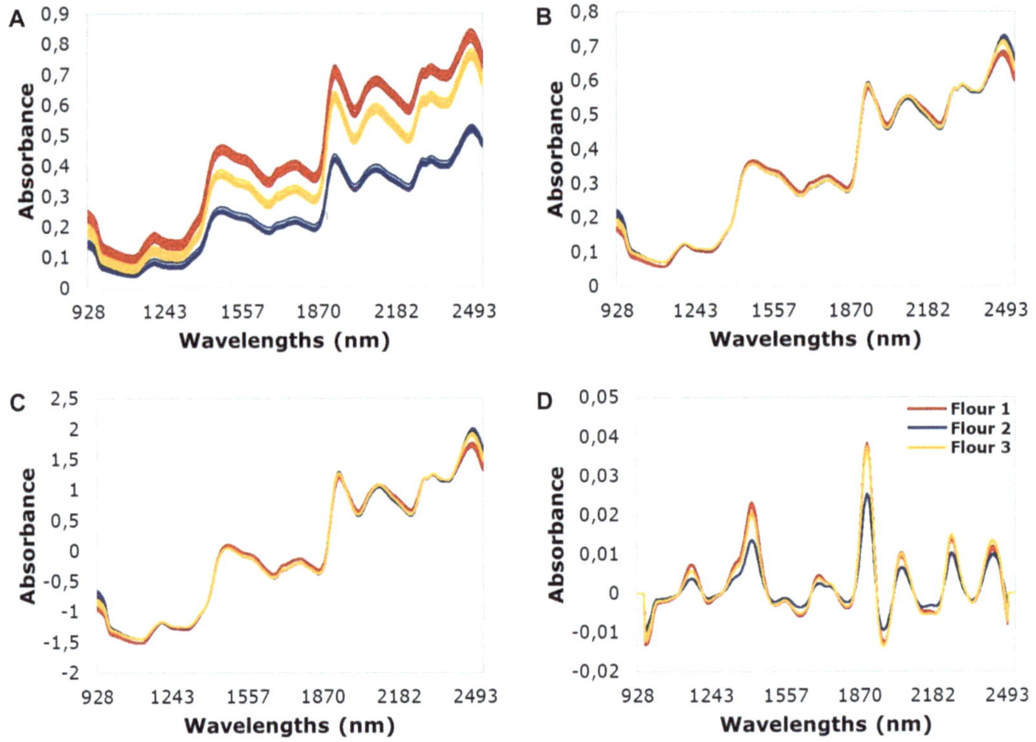

Fig. 2 Spectral information of flour samples: (**A**) raw data, (**B**) MSC, (**C**) SNV, and (**D**) 1st derivative. Source: Authors' collection, 2025

3.4.5 Outliers

1. Usually, there are samples that differ from the general clusters of the dataset, called outliers. This anomaly can be caused by many factors, as instrument failure or experimental conditions.
2. Outliers must be identified, for example, by observing Hotelling's T^2 and Q residuals, and removed from the dataset in order not to affect the accuracy and robustness of the models.

3.4.6 Data Analysis

1. The first step is to perform an exploratory analysis in order to evaluate the behavior of the samples according to the spectral information. The most common and used method to explore the data is principal component analysis (PCA). PCA is an unprovisioned method that reduces the dimensionality of the data into principal components (PC), explaining the variability in the data.
2. Among the plots generated by a PCA, two very important are the scores plot (Fig. 3), which provides a visualization of how samples varied, and the loadings plot (Fig. 4), which shows the important absorption peaks through the spectral range.

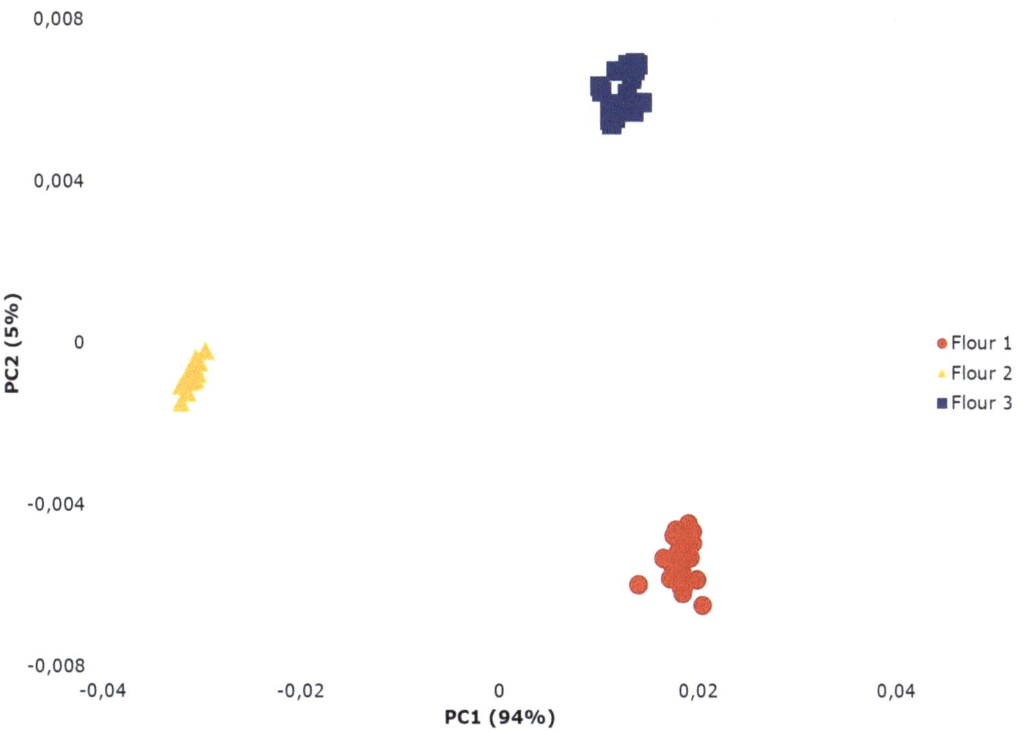

Fig. 3 PCA scores plot of flour spectral data. Source: Authors' collection, 2025

3. After exploratory analysis, calibration models can be developed. There are two main calibration models that can be developed: classification and regression.
4. For classification, it is necessary to define the classes of reference, i.e., what type of categories the samples should belong to. For instance, it could be desired to identify samples with high content or low content of starch. In this case, discriminant models such as linear discriminant analysis (LDA), partial least squares discriminant analysis (PLS-DA), among others, should be used. These models assign the samples to one of the established classes. However, according to the type of analysis, it is possible that the sample does not belong to any of these classes. In this case, a model called soft independent modeling of class analogies (SIMCA) can be used. In this model, a sample can be assigned to one of the classes, to both of the classes or to any of them.
5. For regression analysis (prediction), it is necessary to have a reference value for each sample. For instance, it could be desired to quantify the amount of starch in the samples. Thus, models like partial least squares regression (PLSR), support vector machines (SVM), among others, can be used.

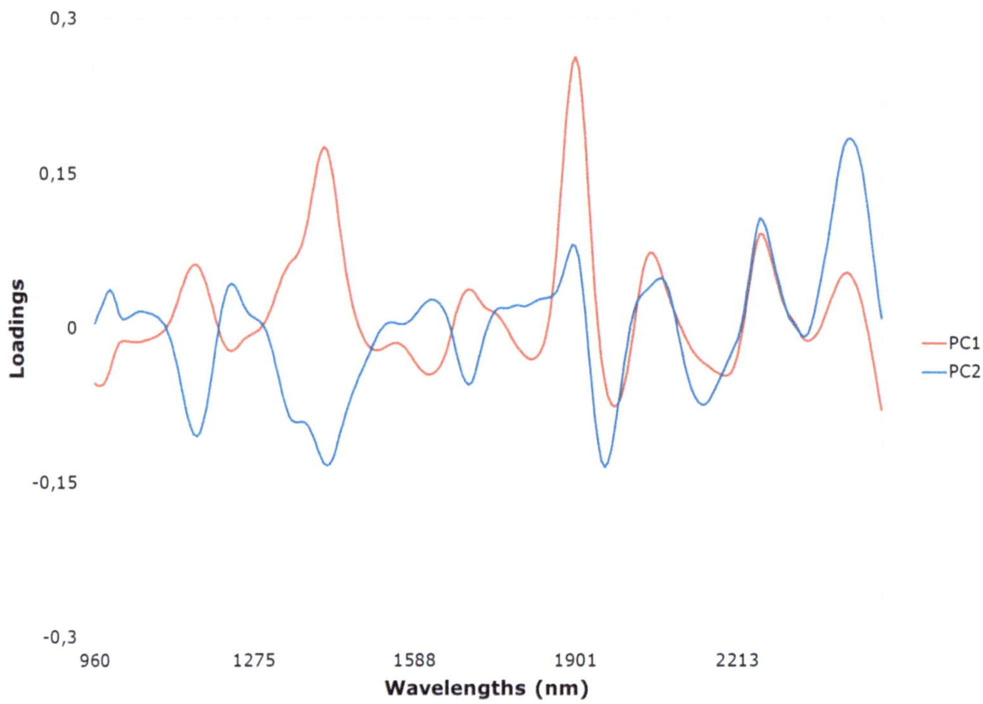

Fig. 4 PCA loadings plot of PC 1 and PCA 2 of flour spectral data. Source: Authors' collection, 2025

3.4.7 Variable Selection

1. Variable selection is an alternative to remove the variables with little or no information in the models, allowing the development of reduced models. This step can also improve the accuracy of the models and reduce de error.
2. It can be carried out manually, by selecting the important wavelengths (peaks) in the loadings plot (Fig. 4), or automatically, by using algorithms as forward stepwise selection (FSS) or Kennard–Stone (KS).

3.4.8 Figures of Merit

1. After the development of the models, the figures of merit, according to each type of calibration (classification or regression), must be evaluated to know the robustness of the models.
2. For classification models, the most important figures of merit evaluated are: accuracy (close to 100%), sensitivity (close to 1), and specificity (close to 1).
3. For regression models, the most important figures of merit evaluated are: coefficient of determination (R^2) of cross-validation and prediction (close to 1), root mean square error

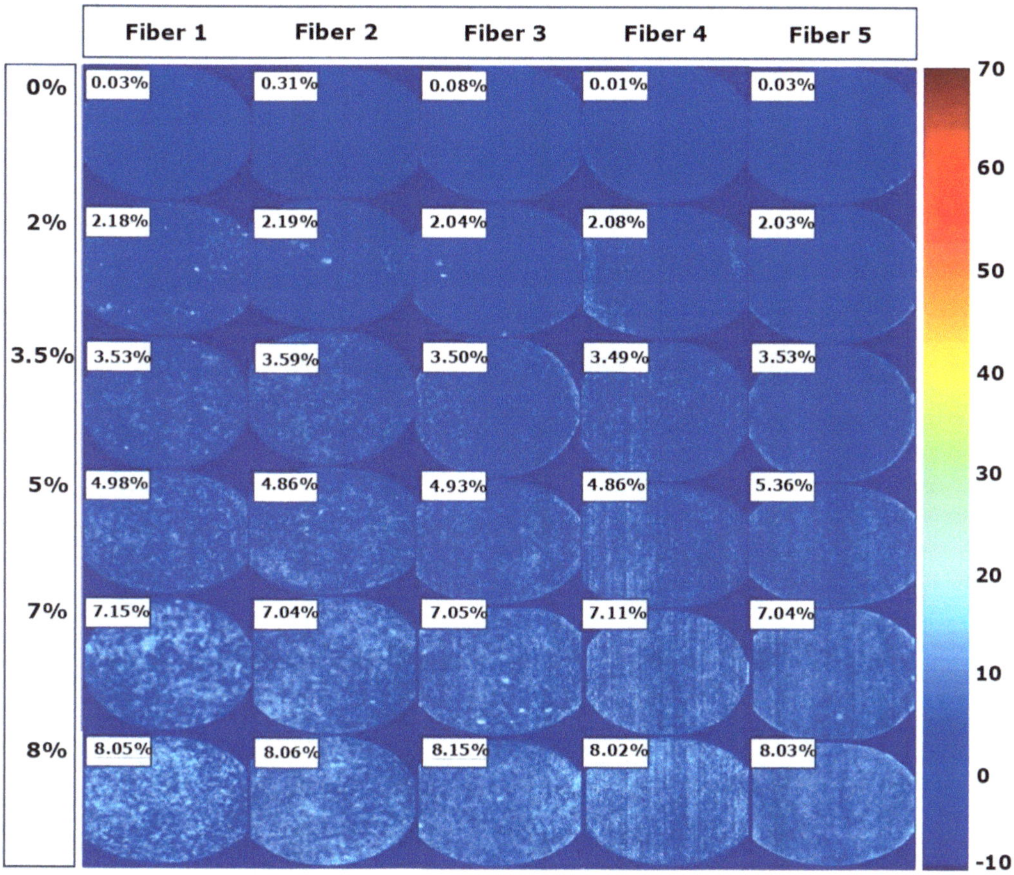

Fig. 5 Distribution maps for visual identification of fiber in flour. The dark blue image represents pure semolina, otherwise fiber. Reprinted from "Identification of fiber added to semolina by near infrared (NIR) spectral techniques," by [7] Food Chemistry, 289, 202.

(RMSE) of cross-validation and prediction (close to 0), and ratio of performance to deviation (RPD) (higher than 3).

3.4.9 Distribution Maps

1. In the case of hyperspectral imaging, the technique allows for the identification of the spatial distribution of the components in the sample. The models with the best accuracy are applied to each pixel of the image for spatial visualization. Figures 5 and 6 show distribution maps developed in order to visualize fiber distribution in flour and pasta. The distribution map of Fig. 5 was developed with mean spectra, while in Fig. 6, the model was applied pixel by pixel on the image.

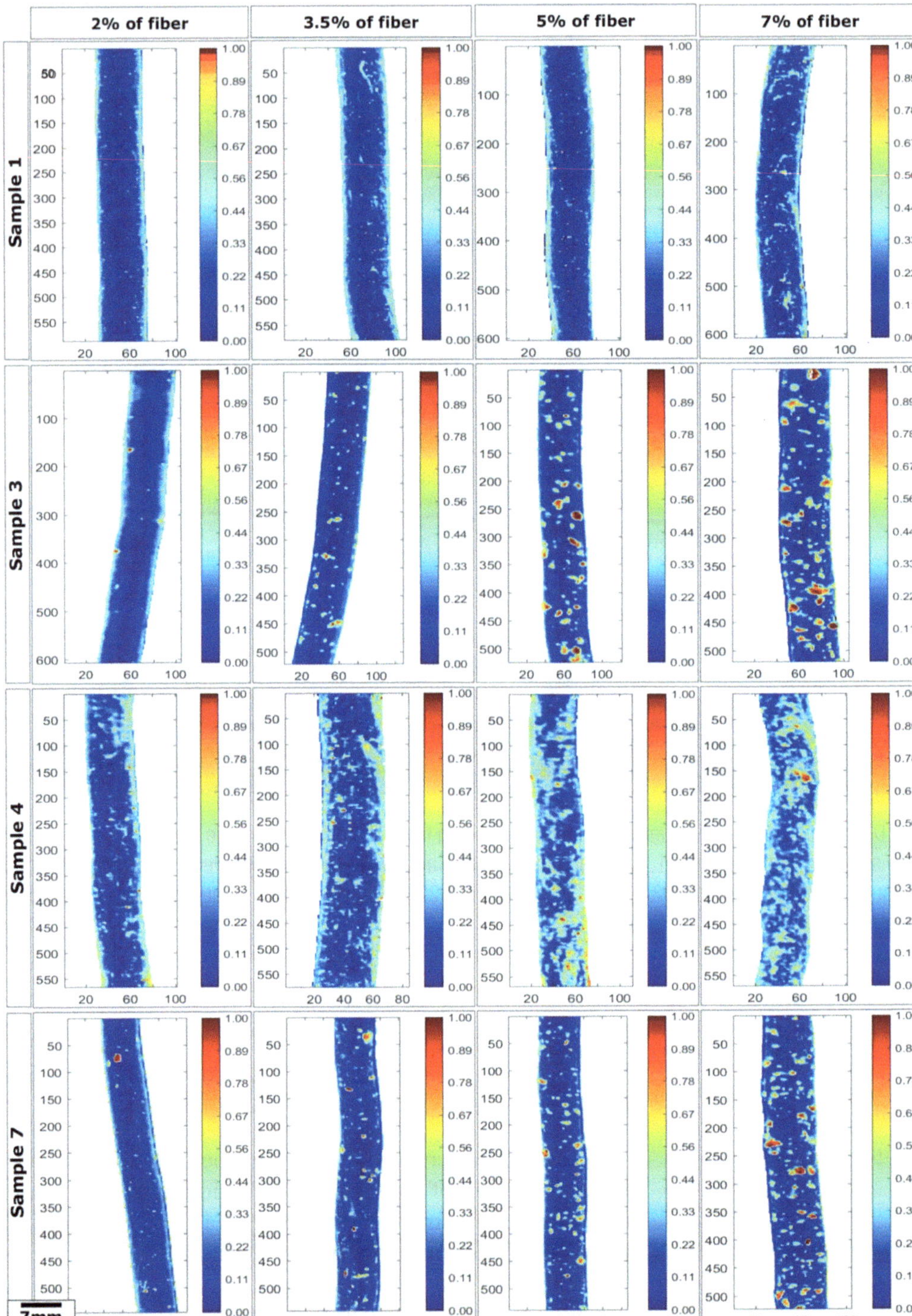

Fig. 6 Distribution maps for visual identification of fiber in pasta. The color represents the percentage of fiber in each pixel of the image. Reprinted from "Near infrared hyperspectral imaging and spectral unmixing methods for evaluation of fiber distribution in enriched pasta," by [8] Food Chemistry, 343, 3. Copyright 2021 Elsevier Ltd. All rights reserved. Reprinted with permission

4 Notes

1. Differences in particle size affect the spectral signal.
2. Samples poorly mixed may not provide a spectrum that represents the whole sample.
3. A Spectralon® gives the highest diffuse reflectance (>99%) of any known material in the UV–VIS–NIR region of the spectrum.
4. Spectra are strongly affected by external lights, and, usually, the experimental analysis is not carried out on the same day. Therefore, in order to minimize these effects, a white reference must be used to correct the sample spectrum or image.

Acknowledgments

This work was supported by the Coordenação de Aperfeiçoamento de Pessoal de Nível Superior—Brasil (CAPES) [Finance Code 001] and São Paulo Research Foundation (FAPESP) [grant numbers 2015/24351-2, 2017/17628-3].

References

1. Peris-Tortajada M (2018) Measuring starch in food. In: Starch in food: structure, function and applications, 2nd edn. Elsevier Ltd
2. Yang C, Zhong F, Douglas GH, Li Y (2019) Study on starch-protein interactions and their effects on physicochemical and digestible properties of the blends. Food Chem 280:51–58
3. AACC International (2011) Approved methods of analysis, 11th edn. http://methods.aaccnet.org/. Accessed 1 Apr
4. AOAC International (2021) Official methods of analysis. AOAC. https://www.aoac.org/. Accessed 1 Apr
5. Porep JU, Kammerer DR, Carle R (2015) On-line application of near infrared (NIR) spectroscopy in food production. Trends Food Sci Technol 46(2):211–230
6. Piqueras S, Burger J, Tauler R, De Juan A (2012) Relevant aspects of quantification and sample heterogeneity in hyperspectral image resolution. Chemom Intell Lab Syst 117:169–182
7. Badaró AT, Morimitsu FL, Ferreira AR (2019) Clerici MTPS, Barbin DF identification of fiber added to semolina by near infrared (NIR) spectral techniques. Food Chem 289: 195–203
8. Badaró AT, Amigo JM, Blasco J, Aleixos N, Ferreira AR, Clerici MTPS, Barbin DF (2021) Near infrared hyperspectral imaging and spectral unmixing methods for evaluation of fiber distribution in enriched pasta. Food Chem 343:128517

Chapter 10

Determining the Molecular Size Distribution of Starch

Dâmaris Carvalho Lima, Bianca Chieregato Maniglia, Meliza Lindsay Rojas and Pedro Esteves Duarte Augusto

Abstract

Starch is composed mainly of two polymers, amylose and amylopectin, which show distinct structures. Amylose is essentially a linear polymer with a lower size, which is formed by α-1–4 glycosidic bonds with few α-1–6 branching points, and this polymer shows a molecular mass between 10^5 and 10^6 g mol^{-1}. Amylopectin is a branched polymer with a larger size, which is formed by α-1–4 glycosidic bonds with a large number of α-1–6 bonds, and this polymer shows a molecular weight from 10^7 to 10^9 g mol^{-1}. These structural differences have an important effect on the characteristics of starch, such as texture, solubility, viscosity, stability, gelatinization, moisture retention, retrogradation, and syneresis, which are the main criteria of selecting an appropriate starch for food products. Moreover, modification processes can partially depolymerize the molecules or promote crosslinking, thus affecting the starch molecular size distribution. In this context, the characterization of the molecular size distribution of starches is important for establishing structure-processing-property relationships, for improvements in human and animal nutrition and industrial applications. In this chapter, we describe the methodology for determining the molecular size distribution of starch using gel permeation chromatography (GPC). GPC separates the starch polymers considering their sizes, which are quantified by separating fractions and using the blue value and/ or the total carbohydrates methods.

Keywords Molecular size, Molecular length, Amylose, Amylopectin, Gel permeation chromatography, Blue value, Total carbohydrates

1 Introduction

Starch is a polysaccharide synthesized by plants and composed mainly of two polymers: amylose and amylopectin. Amylose shows α-D glucose monomers linked by α-1,4-glycosidic bonds, forming linear chains, and this polymer shows a molar mass of ~10^5–10^6 g mol^{-1} [1–3]. Amylopectin is a highly branched molecule with a large number of short-chain branches consisting of 18–31 glucose residues connected by α-1,4-glucoside bonds and side chains with α-1,6 bonds at branch points [4]. The amylopectin polymer shows a very high molar mass (~10^7–10^9 g mol^{-1}) [1–3]. Most starches generally contain

Maria Teresa Pedrosa Silva Clerici and Pedro H. Campelo (eds.), *Basic Protocols in Starch Characterization*, Methods and Protocols in Food Science, https://doi.org/10.1007/978-1-0716-5116-2_10,

about 25%–28% amylose and 72%–75% amylopectin, but depending on botanical sources, genetic modification, climatic growth conditions, and geographic location for cultivation and soil type, the proportion between amylose and amylopectin polymers can vary [5].

These structural differences have an important effect on the characteristics of starch such as texture, solubility, viscosity, stability, gelatinization, moisture retention, retrogradation, and syneresis, which are the main criteria of selecting an appropriate starch for food products [6]. In this context, due to the nutritional and industrial importance of starch, it is indispensable for better knowledge regarding its application and knowledge of its structure [7].

Moreover, modification processes are widely used to enhance starch performance, which can partially depolymerize the molecules or promote crosslinking, thus affecting the starch molecular size distribution.

Gel permeation chromatography (GPC) or also known as size exclusion chromatography (SEC), is a type of chromatography whose separation mechanism is the molecular size, or the hydrodynamic volume of the polymer in solution, with no physical or chemical interaction between the eluents/ samples and the column packaging [8]. This technique consists of using a gel permeation chromatographic column, where the polymer is dissolved and injected in a continuous flow of solvent (mobile phase). The mobile phase flows through porous and rigid particles (the stationary phase) compacted in the column [9]. The column packing works as a molecular filter, and the porous size can be available in a range of sizes [9]. The mechanism of separation consists of the fact that the larger size molecules will not fit into the smaller pores, while the smaller molecules will fit into most of the pores; consequently, small size molecules are retained longer than bigger ones [10]. Therefore, larger molecules elute faster in the column, being collected in the first fractions, while smaller molecules are more retained in the column and take longer to elute, being collected in the later fractions. Figure 1 shows the mechanism of separation involved in the GPC .

The GPC technique enables different fractions of starch molecules to be collected, and each fraction presents the molecules separated in relation to their sizes. To measure the intensity of the molecules present in each fraction, spectrophotometric methods are then used.

The blue value (BV) method is based on the absorbance measurement of the starch–iodine complex, as it is based on the helical structure of amylose's ability to form iodine inclusion complexes, presenting a characteristic blue color with maximum absorption at the 620 nm wavelength [11]. Moreover, in this method, iodine can also form complexes with long linear branched chains of the amylopectin [12, 13]. In this way, it is possible to identify chains of amylose and amylopectin in a BV profile, but these polymers can be distinguished

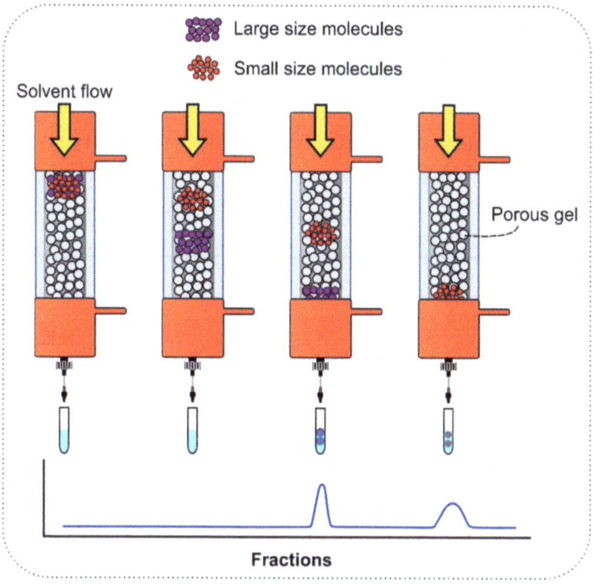

Fig. 1 Schematic representation of the mechanism of separation of the gel permeation chromatography (GPC). Source: Authors' collection, 2025

once the starch molecules are separated by their sizes in the GPC. Therefore, the first peaks represent the large-sized starch molecules (usually amylopectin), which elute first, while the second peak represents the linear and small-sized molecules (usually amylose).

The total carbohydrates (TCH) method, also known as the phenol–sulfuric acid method, detects all classes of carbohydrates, including mono-, di-, oligo-, and polysaccharides [14]. In this method, concentrated sulfuric acid breaks down any polysaccharides, oligosaccharides, and disaccharides into monosaccharides. These compounds then react with phenol to produce a gold–yellow color, and the absorption of which is measured at 490 nm [14]. In this method, it is also possible to identify linear or branched molecules: according to Charles et al. [15], if the intensity of the first peak of the TCH profile is higher than the intensity of the first peak of the BV profile (both compared at the same fraction), it indicates that the amylopectin molecules have shorter branched chains. Moreover, if the TCH profile presents lower peaks than the BV profile, it indicates slightly branched molecules.

Therefore, using one or both methods, it is possible to measure the intensity of the starch molecules in each fraction obtained from the GPC , and thus generate a curve of molecular size distribution. Furthermore, it is important to highlight that GPC is a qualitative assay since there is no quantitative measure of the molecular chain lengths. Even so, it is a widely accepted and applied technique, as well as a valuable tool to understand starch structure.

In this chapter, we describe the methodology for determining the molecular size distribution of starch using gel permeation chromatography to make a separation of the starch molecules considering their sizes and then use the blue value and total carbohydrates methods to quantify the intensity of the starch molecules in each fraction.

2 Material

It is recommended that the solutions be prepared using distilled water or deionized water. All the solutions can be prepared at room temperature (~25°C).

2.1 Solutions Preparation

1. Ethanol solution (20% v/v): To prepare 1 L of this solution, add 208 mL of ethanol (96%) in a volumetric flask (1 L) and complete with deionized water to the meniscus.
2. Eluent solution (containing 25 mM NaCl and 1 mM NaOH): In a beaker, dissolve 7.2 g of NaCl in 50 mL of deionized water, and then, in another beaker, dissolve 0.2 g of NaOH in 50 mL of deionized water. After this, mix the NaCl solution with the NaOH solution. Then add them to a volumetric flask (5 L), and then complete with deionized water to the meniscus.
3. Dimethyl sulfoxide solution (DMSO) solution (90% v/v): To prepare 1 L of this reagent, add 900 mL of dimethyl sulfoxide (anhydrous, ≥99.9%) in a volumetric flask (1 L) and complete with deionized water to the meniscus.
4. Acetate buffer 0.2 M solution at pH 5.0: To prepare 1 L of this reagent, add 27.2 g of sodium acetate trihydrate and 6 mL of glacial acetic acid to a volumetric flask (1 L), and then complete with deionized water to the meniscus.
5. I_2/KI stock solution: Weigh 1 g of potassium iodide (KI) and 0.1 g of iodine (I_2), dissolve them in 20 mL of the acetate buffer 0.2 M solution at pH 5.0. After this, add the solution to a volumetric flask (100 mL) and complete with acetate buffer (0.2 M solution at pH 5.0) to the meniscus. Store the stock solution in an amber glass bottle at 4°C. Dilute this stock solution before using it in the samples, as follows: add 10 mL of I_2/KI stock solution to 90 mL of deionized water (*see* **Note 1**).
6. Phenol solution (5%): Weigh 5 g of phenol (P.A.) and dissolve it in 50 mL of deionized water. After this, add the solution to a volumetric flask (1 L) and complete it with deionized water.

3 Method

3.1 Gel Permeation Chromatography (GPC) Column Preparation

1. To pack the GPC column, use the Sepharose CL-2B gel.
2. Before packing the GPC column, calculate the amount of gel slurry needed to pack the column, such as follows:

$$S_v = \frac{C_v - C_f}{S_c}$$

Where

S_v = Slurry volume (mL); C_v = Column volume (mL); C_f = Compression factor; and S_c = Slurry concentration.

The information about the C_f and S_c can be found in the user instructions of the Sepharose CL-2B gel.

3. Before proceeding to packing, it is necessary to prepare the gel slurry: mix 75% of the Sepharose CL-2B gel with 25% of the eluent solution or ethanol solution (20% v/v) (*see* **Note 2**).
4. Degas the gel slurry before packing (*see* **Note 3**).
5. All material that will be used for packing must be equilibrated to the temperature at which the chromatography will be performed (in general, at room temperature or using an air conditioning system).
6. Put the column (without the bottom and top adapters) in vertical position on a stable laboratory stand (*see* **Note 4**).
7. Before inserting the bottom adapter in the column, remove all traces of air in it by forcing eluent or ethanol solution (20% v/v) up through the net (*see* **Note 2**, using a syringe).
8. Leave a thin layer of eluent or ethanol solution (20% v/v) over the net of the bottom adapter and insert it into the column.
9. Insert with the syringe more eluent or ethanol solution (20% v/v) in the column through the bottom adapter. The column must be filled with approximately 1 cm of the eluent or ethanol solution (20% v/v); this is recommended to avoid the formation of bubbles during gel slurry packing.
10. Close the column bottom outlet with a stop plug.
11. Pour the gel slurry into the column reservoir (all the gel slurry should be poured in a single operation to avoid the formation of bubbles) (*see* **Note 5**).
12. If the gel slurry does not reach the top of the column, fill it to the top with the eluent or ethanol solution (20% v/v).
13. Remove the stop plugin of the bottom adapter outlet and put the top adapter in the column.

14. Insert with a syringe eluent or ethanol solution (20% v/v) in the column through the top adapter to remove all traces of air in this part of the column.
15. To finalize the column packaging, connect the top adapter to the peristaltic pump and turn it on to start pumping eluent (descending mode) at a flow rate recommended by the user instructions of the Sepharose CL-2B gel (*see* **Note 6**).
16. Pack the gel until the gel bed has reached a constant height and stop the pump.
17. Adjust the mobile part of the top adapter a further 2–4 mm on the gel surface and then let the packed column rest for about 24 h.
18. After about 24 h, use a flashlight in the packed column to check if it does not present any cracks, fissures or bubbles. If the packed column does not present any cracks, fissures or bubbles, the column is ready to use. On the contrary, if any of these problems occur (*see* **Note 7**).
19. To sample run, insert the tubing of the bottom adapter on the sample solution and connect the top adapter to a fraction collector and start to pump (ascending flow) (*see* **Note 4**). After running the sample, stop the pump. Remove the tubing from the sample solution and clean the external area of the tubing (to remove the sample from it). Then, insert the tubing in the eluent solution (to elute the sample solution through the column) and start to pump again (*see* **Note 6**). Figure 2 shows a representation of the GPC installed column following the steps described here.

3.2 Sample Preparation

1. Weigh 0.1 g (on a dry basis) of starch and disperse it in 10 mL of dimethyl sulfoxide solution (90%).
2. Place the starch solution in a boiling water bath (~100°C) under constant stirring for 1 h (*see* **Note 8**).
3. Place the starch solution in a water bath at 25°C under constant stirring for 16 h.
4. Mix an aliquot of 3 mL of the starch solution with 10 mL of absolute ethanol.
5. Centrifuge the solution at 3000 × *g* for 30 min (*see* **Note 9**).
6. Discard the supernatant, redissolve the precipitated starch in 9 mL of boiling water, and then add 1 mL of glucose (1 mg/ mL) (*see* **Note 10**).
7. Place the precipitated starch with the glucose in a boiling water bath (~100°C) with constant stirring for 30 min.
8. Cool the obtained solution at room temperature (until ~25°C) and then filter (5 μm) it to remove the insoluble residues (*see* **Note 11**).

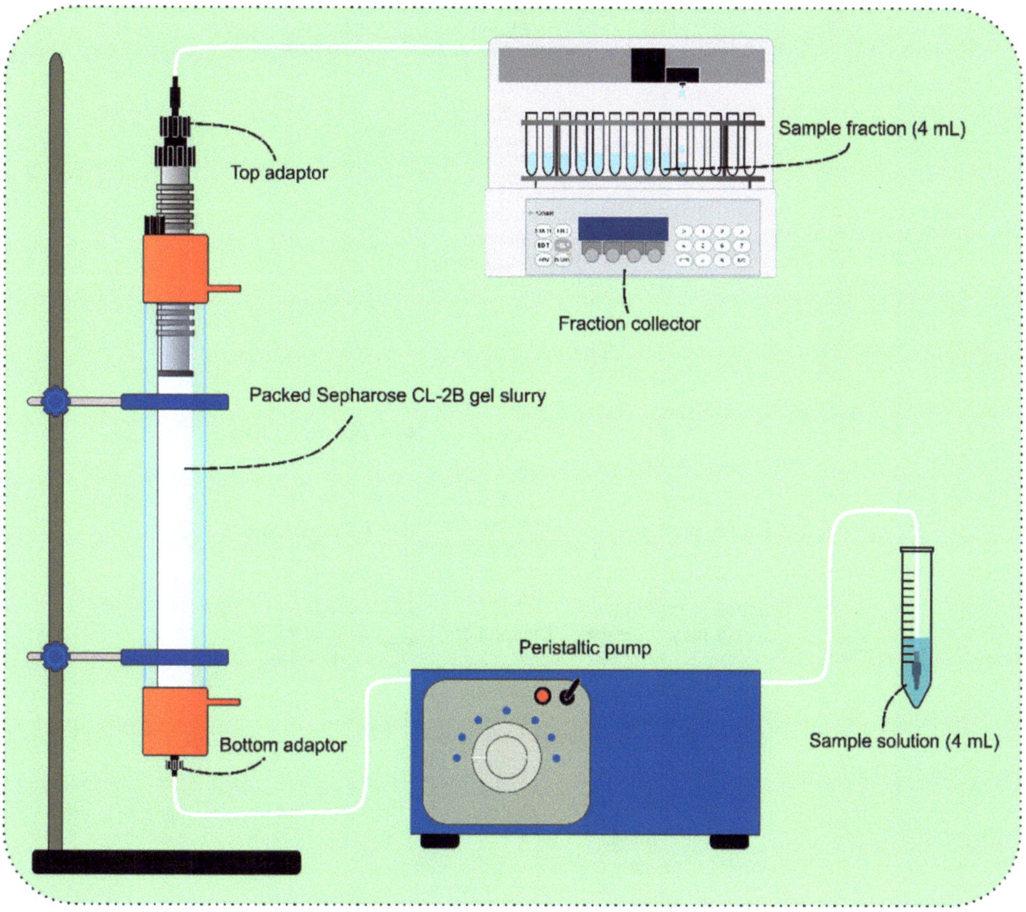

Fig. 2 Gel permeation chromatography analysis. Source: Authors' collection, 2025

9. Inject 4 mL of the obtained solution into the GPC column.
10. Use the eluent solution to elute the sample through the column (*see* **Note 12**).
11. Run the column by using a peristaltic pump, with the eluent solution flowing in the ascending mode.
12. During the sample run, collect the sample fractions (aliquots of eluent solution + starch solution) released from the column using a fraction collector (*see* **Note 13**), programmed to collect a certain volume of sample per tube (for example, 4 mL per tube) (*see* **Note 14**). The total amount of small test tubes necessary to collect the fractions will depend on the volume per tube. For instance, for 4 mL of sample volume, will be necessary about 280 small test tubes to collect the sample fractions. Figure 3 shows the flowchart of sample preparation for GPC analysis.
13. After collecting the sample fractions, use them to carry out the BV and TCH analysis.

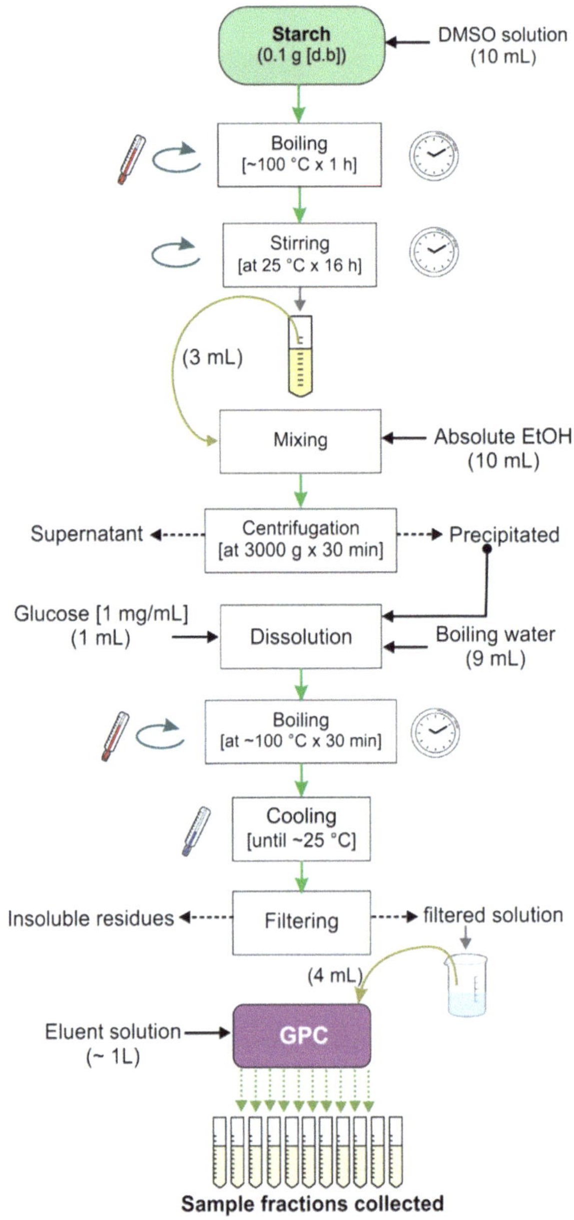

Fig. 3 Flowchart of sample preparation for GPC analysis and obtention of sample fractions for BV and TCH analyses. Source: Authors' collection, 2025

3.3 Blue Value (BV) Analysis: Molecular Size Distribution of Amylopectin/Amylose Molecules

1. Mix 200 μL of the sample solution with 400 μL of I_2/KI stock solution in a test tube. Then, add 3 mL of deionized water and stir the solution.
2. After that, analyze the samples in a spectrophotometer at a wavelength of 620 nm (*see* **Note 15**). Then, record the absorbance values of each fraction and use them to make a graph (absorbance at 630 nm vs. fractions).

3. Figure 4 displays an example of the BV profile of native cassava starch [16]. It is worth highlighting that this profile can be obtained with different quantities of fractions, once it is possible to alter the volume collected in each tube during the chromatographic separation (step 12, described in Sect. 3.2). In the BV profile, the first fraction peak represents the bigger molecules of the starch (such as amylopectin), whereas the second fraction peak represents the smaller molecules (such as amylose).

3.4 Total Carbohydrates (TCH) Analysis: Molecular Size Distribution of Total Carbohydrates

1. Mix 1 mL of the sample solution with 1 mL of phenol solution (5%) in a test tube.
2. Add 5 mL of concentrated sulfuric acid (95%) to the sample solution (*see* **Note 16**). Then, stir the samples and maintain them at rest for 10 min.
3. Place the samples in a water bath at 30°C under constant stirring for 20 min.
4. Cool the samples at room temperature (until ~25°C).
5. Analyze the samples in a spectrophotometer at a wavelength of 490 nm (*see* **Note 15**). Then, record the absorbance values and use them to make a graph (absorbance at 490 nm vs. fractions).
6. Figure 4 displays an example of the total carbohydrates (TCH) profile of a native cassava starch [17].
7. As mentioned in step 4 of the BV analysis, the number of fractions can vary. In the TCH profile, the first fraction peak represents the presence of bigger carbohydrate molecules (such as amylopectin), whereas the second fraction peak represents the presence of smaller carbohydrate molecules (such as amylose).

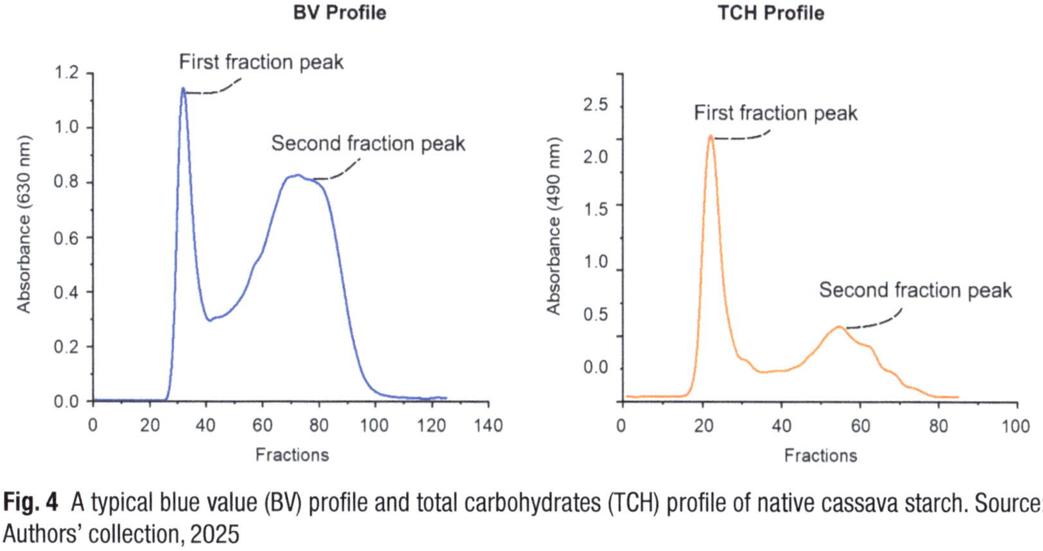

Fig. 4 A typical blue value (BV) profile and total carbohydrates (TCH) profile of native cassava starch. Source: Authors' collection, 2025

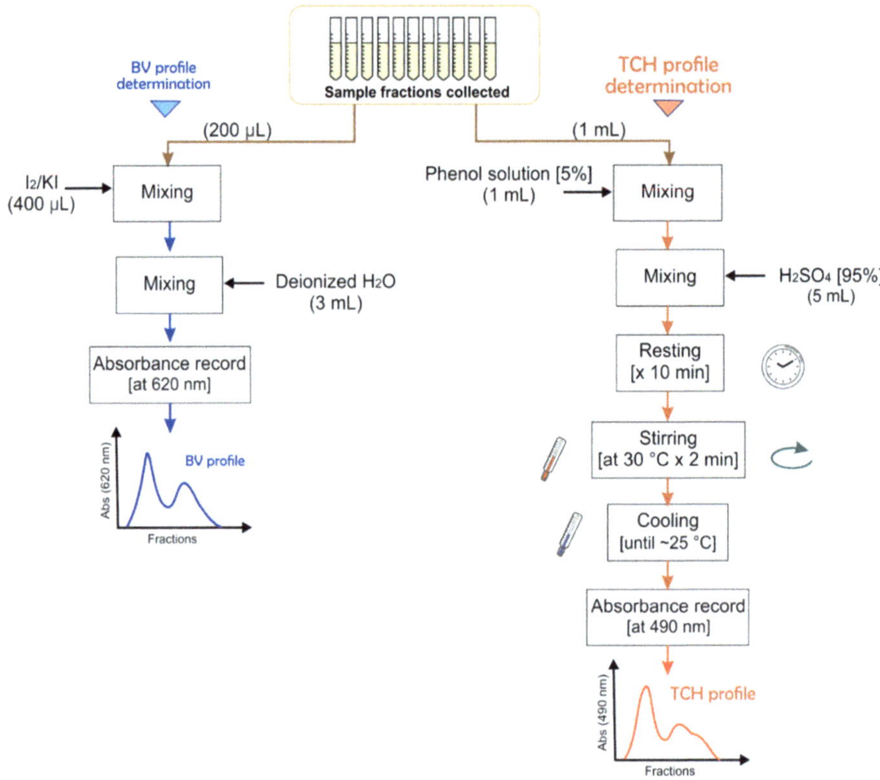

Fig. 5 Flowcharts of BV and TCH procedures. Source: Authors' collection, 2025

As mentioned, the comparison between the intensity of the TCH and BV peaks brings information about the presence of branched chains in the amylopectin polymer [13].

Figure 5 shows flowcharts of the BV and TCH procedures.

4 Notes

1. Instead of preparing an I_2/KI solution, this ready-made solution can be purchased under the name of Lugol solution.
2. Ethanol solution (20% v/v) is better than the eluent solution once it breaks surface tensions.
3. Use a vacuum filter flask and a vacuum pump to degas the gel. Carry out this procedure as reported in the user instructions. This step is important because it reduces the chance of forming air bubbles in the packed column.

4. During the packing process of the gel and while running the samples in the column, avoid places that are exposed to direct sunlight. It is also important to pack the column and run the samples in a place that has a constant temperature.
5. Use a glass rod or spatula held against the wall of the column to pour the gel slurry to minimize the formation of air bubbles.
6. In column packaging, the flow direction of the eluent is descending, and for the sample run, the eluent presents ascending flow. In this way, the inlet and the outlet of the column change if it is in packaging or sample run way.
7. If the column presents any cracks, fissures or bubbles, the packaging process must be remade. Therefore, it is necessary to remove the gel from the column and repeat steps from 6 until 18.
8. The beaker in which the sample was placed must be capped to prevent loss of moisture.
9. Instead of filtering the samples, they can be centrifuged at 3000 × *g* for 30 min, and the next step is done with the supernatant.
10. The glucose is used as a marker to indicate the end of the analysis.
11. In this step, the samples can be stored for seven days at ~5°C for further analysis.
12. The eluent volume is particular to each starch sample. Using the BV and TCH procedures is confirmed if all samples run through the column.
13. This step can be done manually instead of using a fraction collector. However, carrying out this step manually is not so accurate, and it is more laborious than using a fraction collector.
14. If one wanted a broader curve profile, reduce the fraction volume of the sample per tube. This reduction of the fraction volume will require more small test tubes to collect fractions.
15. The samples can also be analyzed in a microplate reader at the same wavelength as the spectrophotometer. The advantage of using a microplate reader is that one can measure several samples at once, using less reagent, and the final volume for the analysis can be reduced. The sample volume for analysis using cuvettes for a spectrophotometer has to be at a minimum of 3 mL, while 0.2 mL is generally enough for the microplate reader. In this way, the quantity of the sample and the reagents can be reduced, but following the proportions mentioned in the methods.

16. In order to promote good homogenization of the acid/ sample, place the concentrated sulfuric acid (95%) directly in the sample solution, not letting the acid slide down the tube walls.

5 Literature Results

Figure 6 shows two different works that used the same methodology of molecular size distribution determination described in this chapter. Figure 6A shows a blue value and total carbohydrates profiles were obtained for different native starch sources: arracacha morada, arracacha amarilla A, arracacha amarilla B, maize, waxy, and high amylose [18]. Figure 6B shows a blue value and total carbohydrates profiles obtained for cassava starch native and modified by different methods (ozone treatment—O_3, dry heating treatment—DHT, and their combinations) [17].

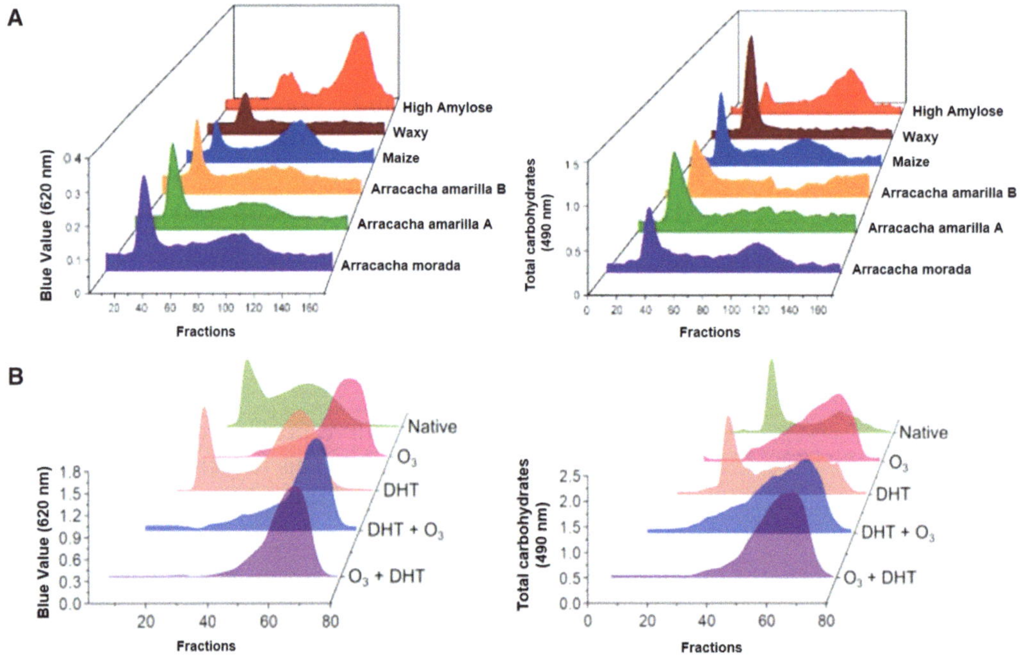

Fig. 6 Blue Value and total carbohydrates profiles of **A** samples of different nativenative starches sources (arracacha morada, arracacha amarilla A, arracacha amarilla B, maize, waxy, and high amylose) [18], and **B** cassava starch native and modifiedmodified by different methods isolated or combined (ozone treatment—0_3, dry heating treatment—DHT) [17]. Reprinted with permission from Castanha et al. [18] and Lima et al. [17]. Copyright 2018 and 2021, with permission from Elsevier

Acknowledgments

The authors are grateful to São Paulo Research Foundation (FAPESP, Brazil) for funding the project n° 2019/05043-6 and the BC Maniglia Young Investigator grant (2021/05947-2); the "Coordenação de Aperfeiçoamento de Pessoal de Nível Superior—Brasil (CAPES, Brazil)"—Finance Code 88882.378469/2019-01, for financing the Dâmaris C Lima Ph.D. scholarship. The authors are also grateful to Dr. Manoel Divino Matta Junior for general support. Communauté urbaine du Grand Reims, Département de la Marne, Région Grand Est, and European Union (FEDER Champagne-Ardenne 2014–2020, FEDER Grand Est 2021–2027) are acknowledged for their financial support to the Chair of Biotechnology of Centrale Supélec and the Centre Européen de Biotechnologie et de Bioéconomie (CEBB).

References

1. Gidley MJ, Hanashiro I, Hani NM et al (2010) Reliable measurements of the size distributions of starch molecules in solution: current dilemmas and recommendations. Carbohydr Polym 79:255–261
2. Jane J-L (2009) Structural Features of Starch Granules II. In: Bemiller JN, Whistler R (eds). Starch: chemistry and technology, 3rd edn. Academic Press, London, pp 193–236
3. Bertoft E (2004) Analysing starch structure. In: Eliasson A-C (ed). Starch in food: structure, function andapplication, 1st edn. Woodhead Publishing, Cambridge/Boca Raton, pp 57–96
4. Wang Y, Chen L, Yang T, Ma Y, McClements DJ, Ren F, Tian Y, Jin Z (2021) A review of structural transformations and properties changes in starch during thermal processing of foods. Food Hydrocoll 113:106543
5. Van Hung P, Maeda T, Morita N (2006) Waxy and high-amylose wheat starches and flours—characteristics, functionality and application. Trends Food Sci Technol 17:448–456
6. Klaochanpong N, Puttanlek C, Rungsardthong V, Puncha-arnon S, Uttapap D (2015) Physicochemical and structural properties of debranched waxy rice, waxy corn and waxy potato starches. Food Hydrocoll 45:218–226
7. Szwengiel A, Kubiak P (2020) Molecular dispersion of starch as a crucial parameter during size-exclusion chromatography. Foods 9:1204
8. Striegel AM (2017) Size-exclusion chromatography. In: Liquid chromatography. Elsevier, pp 245–273 Amsterdan https://doi.org/10.1016/B978-0-323-99968-7.00022-9
9. Payerle G, Team RC, Payerle GDS, Dolnicar S, Chapple A, Pastuszak AW, Wang R (2015) Scholar (3). Ann Tour Res 3:45
10. Salemis P, Rinaudo M (1984) Gel permeation chromatography of starch and other uncharged polysaccharides. Polym Bull 11(4):397–400
11. STANDARDIZATION IOFOR (1987) ISO 6647: Riz-détermination de la teneur en amylose
12. Liu Q (2005) Understanding starches and their role in foods. In: Cui SW (ed). Food carbohydrates. CRC Press, Boca Raton, pp 309–355
13. Shen X, Bertoft E, Zhang G, Hamaker BR (2013) Iodine binding to explore the conformational state of internal chains of amylopectin. Carbohydr Polym 98:778–783
14. Nielsen SS (2003) Phenol-sulfuric acid method for total carbohydrates. In: Food analysis laboratory manual, food science texts series. Springer-Verlag, USA, pp 39–44
15. Charles AL, Chang Y-H, Ko W-C, Sriroth K, Huang T-C (2004) Some physical and chemical properties of starch isolates of Cassava genotypes. Starch 56:413–418
16. Lima DC, Castanha N, Maniglia BC, Matta Junior MD, La Fuente CIA, Augusto PED (2021) Ozone processing of Cassava starch. Ozone Sci Eng 43:60–77
17. Lima DC, Maniglia BC, Matta Junior MD, Le-Bail P, Le-Bail A, Augusto PED (2021) Dual-process of starch modification: combining ozone and dry heating treatments to modify cassava starch structure and functionality. Int J Biol Macromol 167:894–905
18. Castanha N, Villar J, Matta Junior MD, Anjos CBPdos, Augusto PED (2018) Structure and properties of starches from Arracacha (*Arracacia xanthorrhiza*) roots. Int J Biol Macromol 117:1029–1038

Chapter 11

Determining the Carbonyl and Carboxyl Group Content on Starch

Dâmaris Carvalho Lima, Bianca Chieregato Maniglia, Meliza Lindsay Rojas and Pedro Esteves Duarte Augusto

Abstract

Starches can be naturally oxidized during storage by external agents, or they can be intentionally oxidized in order to promote changes in their structure and properties. Starch oxidation occurs mainly in the hydroxyl groups at positions C-2, C-3, and C-6, forming carbonyl and carboxyl groups. In this way, the contents of the carbonyl and carboxyl groups indicate the extent of starch oxidation. In this chapter, we describe methodologies based on titrimetric methods for determining the carbonyl and carboxyl group content on starch samples. In addition, we also discuss some notes to guide the readers in conducting the experiments.

Keywords Oxidation, Oxidized starch, Carbonyl groups, Carboxyl groups, Quantitative methods, Titrimetric methods

1 Introduction

Starch is a polysaccharide synthesized by photosynthesis in the chloroplast and amyloplast of plant cells, being stored as an energetic source for plants [1]. In addition, starch is an important ingredient in the human diet, and it has also been used in many industrial applications, such as in the food (human and animal), pharmaceutical, cosmetic, chemical, petrochemical, and textile sectors, among others [2].

Starches can be oxidized during storage by external agents such as air, light and heat, or they can be intentionally oxidized in order to promote changes in their structure and properties. Oxidized starches can overcome some deficiencies presented by native starches, as they may present better solubility in water [3], excellent filmogenic properties [4], lower viscosity in high solids content, less retrogradation, increased gel strength and better clarity [5] than native starches. Therefore, the oxidation process allows the expansion of the starch potential for industrial applications.

Maria Teresa Pedrosa Silva Clerici and Pedro H. Campelo (eds.), *Basic Protocols in Starch Characterization*, Methods and Protocols in Food Science, https://doi.org/10.1007/978-1-0716-5116-2_11,

The oxidation of starch occurs mainly in the hydroxyl groups at positions C-2, C-3, and C-6 in a d-glucopyranosyl unit, with the formation of carbonyl and carboxyl groups [6]. Figure 1 shows a schematic representation of starch oxidation, including the formation of carbonyl and/or carboxyl groups in the chains of amylose and/or amylopectin molecules. It is worth mentioning that the reaction between the oxidizing agent and starch can also result in partial hydrolysis of the polymer chains, which are then depolymerized by cleavage of glycosidic bonds [6].

In the literature, there are descriptions of the mechanism involved in starch oxidation. Some authors claim that hydroxyl groups are first oxidized to carbonyl groups, and then to carboxyl groups [7, 8]. However, according to Butrim et al. [9], there is the possibility of the formation of carboxyl and carbonyl groups in both routes, consecutive or in parallel. These authors state that depending on the type of

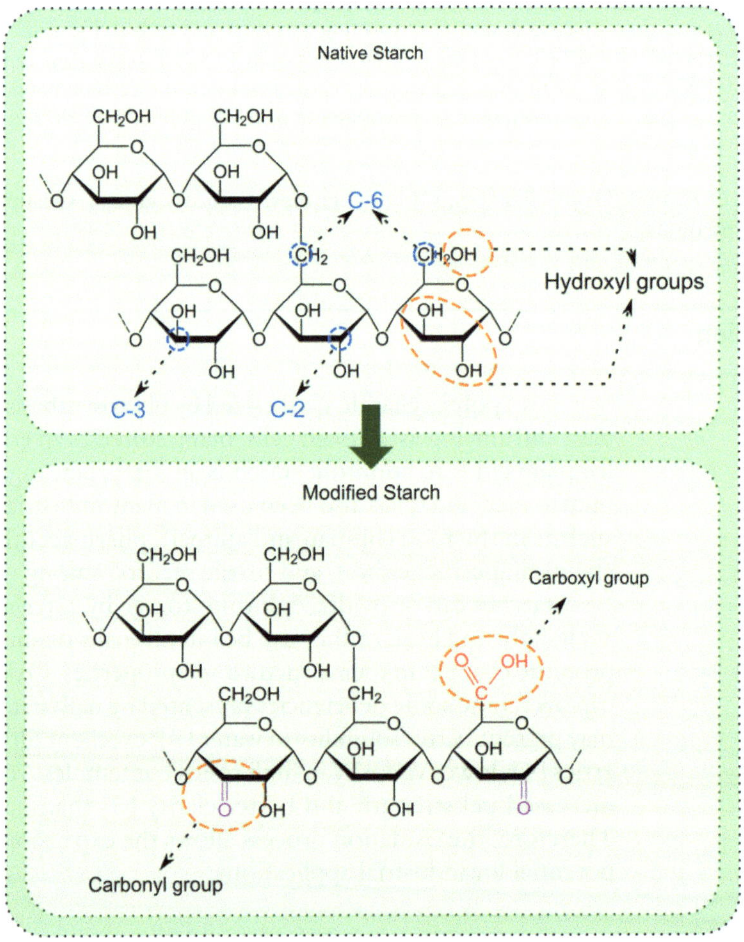

Fig. 1 Schematic representation of starch oxidation: formation of carbonyl and carboxyl groups in the chains of amylose and/or amylopectin molecules. Source: Authors' collection, 2025

oxidant, carbonyl groups are selectively formed by the oxidation of hydroxyl groups at positions C-2 and C-3, while with other oxidants, hydroxyl groups at position C-6 form carboxyl groups.

The level or degree of oxidation of starch can also be determined by the sum of the content of carbonyl and carboxyl groups per glucose unit [10]. Therefore, the higher the value of the content of these new groups, the greater the degree of substitution due to oxidation [11]. Also, the effectiveness of oxidation depends on the type of oxidant used, the botanical origin of the starch, and the conditions of the process [12]. In industrial processes, the most common starch oxidizing agents are sodium hypochlorite and potassium periodate [13]. However, the use of chemical reagents causes concerns related to the final product and the environment, due to the residues formed. For instance, some limitations exist in the food sector. Therefore, there is a great search for green technologies of starch oxidation, that are considered environmentally friendly, such as ozone, cold plasma and dry heating treatment [14, 15].

In the literature, some techniques can qualitatively determine the presence of carbonyl and carboxyl groups, such as infrared spectroscopy [16]. However, for quantitative analysis of these groups, the titrimetric methods developed by Smith [17] for carbonyl content determination and by Chattopadhyay et al. [18] for carboxyl content determination are accepted by different research groups.

In this context, this chapter describes the methodologies based on Smith [17] and Chattopadhyay et al. [18] for determining carbonyl and carboxyl group contents on starch, respectively. Also, we discuss notes to guide readers in carrying out the experiments.

2 Materials

Prepare all solutions using deionized water or distilled water. All the solutions can be prepared at room temperature (~25°C), and they can be stored in the refrigerator.

2.1 Solutions Preparation

1. Hydrochloric acid (HCl) 0.1 M solution: To prepare 1 L of this reagent, add 8.4 mL of HCl (P.A.) in a volumetric flask (1 L) and then complete with deionized water to the meniscus.
2. Sodium hydroxide (NaOH) 0.5 M solution: To prepare 1 L of this reagent, dissolve 20 g of NaOH in 300 mL of deionized water (careful: it is an exothermic process; therefore, use an ice bath to cool the solution). Then, add the solution to a volumetric flask (1 L) and complete with deionized water to the meniscus.
3. Sodium hydroxide (NaOH) 0.01 M solution: To prepare 1 L of this reagent, add 20 mL of the NaOH 0.5 M solution in a volumetric flask (1 L) and complete with deionized water to the meniscus.

4. Hydroxylamine hydrochloride solution: First, 25 g of hydroxylamine is dissolved in 50 mL of deionized water. After this, 100 mL of NaOH 0.5 M solution is added to this solution. Then, add the solution to a volumetric flask (500 mL) and complete with deionized water to the meniscus.

3 Methods

3.1 Determination of Carbonyl Group Content

1. Weigh 4 g (on dry basis, that is, without considering the water contained in the sample) of the starch samples (native or oxidized starch) and then disperse the samples in 100 mL of deionized water in a beaker (*see* **Notes 1 and 2**).
2. Blank: Add 100 mL of deionized water in a beaker (without starch) (*see* **Note 3**).
3. Place the beakers with the samples and the blank in a boiling water bath (~100°C) with constant vigorous stirring, to guarantee suspension, for 30 min (or until complete gelatinization of the starch samples) (*see* **Note 4**).
4. Cool the gelatinized samples and the blank in a water bath until reaching 40°C.
5. Adjust the pH of the samples and the blank to 3.2 using a 0.1 M HCl solution (*see* **Note 5**).
6. Add 15 mL of hydroxylamine hydrochloride solution to all the samples, including the blank.
7. Place the samples and the blank in a water bath at 40°C with constant stirring for 4 h.
8. Titrated the samples and the blank with the HCl 0.1 M solution, until pH 3.2 (*see* **Notes 5 and 6**). Record the total volume titrated to use this value for calculations of carbonyl group content.
9. In order to calculate the carbonyl group content, use the following equation:

$$\frac{\mathrm{CO}}{100\ \mathrm{GU}} = \frac{[(V_{\mathrm{bl}} - V_{\mathrm{ss}}) \cdot M_{\mathrm{HCl}} \cdot 0.028 \cdot 100]}{\mathrm{SM}}$$

Where

V_{bl} = Titration volume (mL) of the HCl 0.1 M solution required for the blank;

V_{ss} = Titration volume (mL) of the HCl 0.1 M solution required for the starch sample;

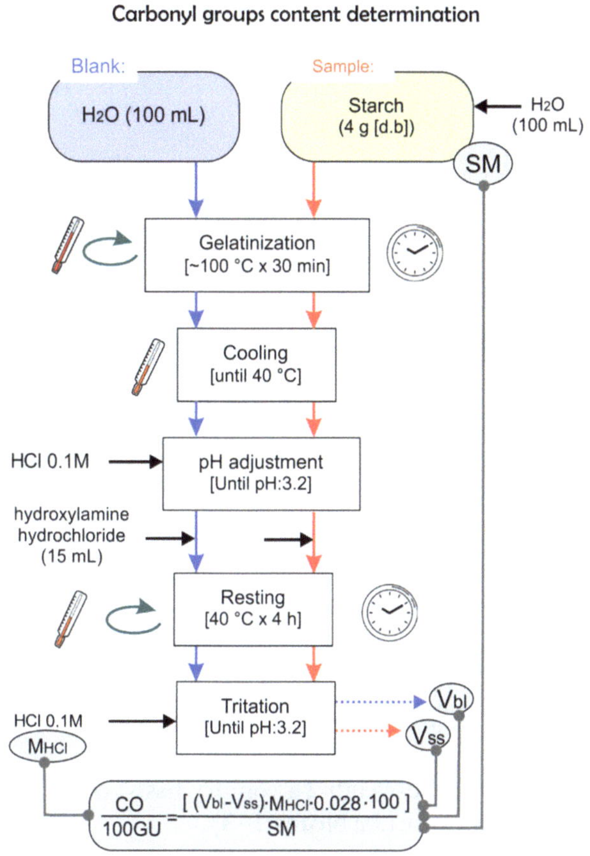

Fig. 2 Flowchart of the procedure to determine the content of carbonyl groups. Source: Authors' collection, 2025

M_{HCl} = Molarity of the HCl solution (mol/L);
0.028 = Molecular weight of carbonyl unit (M(C) + M(O) = 0.028 kg/mol);
SM = Sample mass (g) on dry basis.

10. The result is expressed in quantity of carbonyl groups per 100 glucose units (CO/100 GU).
11. Figure 2 presents the flowchart of this analysis.

3.2 Determination of Carboxyl Group Content

1. Literature expects carboxyl groups to be present only in oxidized starches. Therefore, this assay is conducted only for oxidized starches (after storage or intentionally modified), being the correspondent native starch the blank.

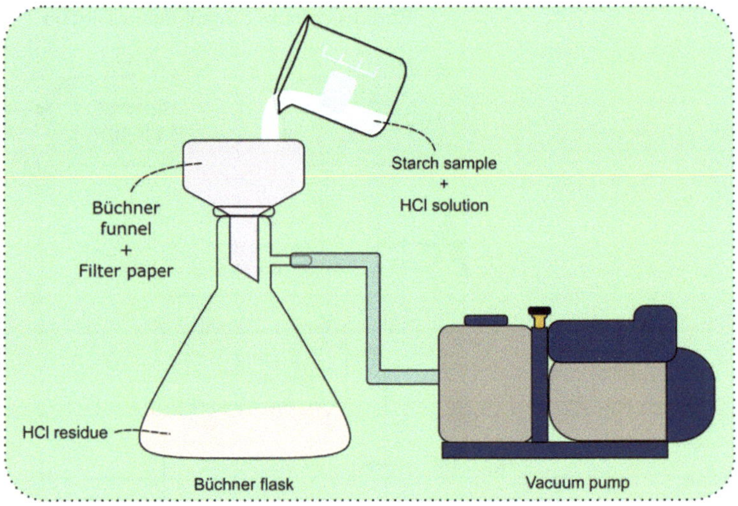

Fig. 3 Filtration process of the starch sample. Source: Authors' collection, 2025

2. Weigh 2 g (on a dry basis) of the oxidized starch sample in a beaker (*see* **Notes 1 and 2**).
3. Blank: weigh 2 g (on dry basis) of the native starch sample in beaker (*see* **Notes 1–3**).
4. Suspend the sample in 25 mL of HCl 0.1 M solution at room temperature (~25°C).
5. Constantly stir the sample with the HCl 0.1 M solution for 30 min.
6. Filter the sample immediately (*see* **Note 7**): Place a filter paper (205 μm) in a Büchner funnel (*see* **Note 8**). Then, place the Büchner funnel with the filter paper in a Büchner flask. After this, gradually place the sample with the HCl 0.1 M solution on the filter paper. In order to make the filtration process faster, use a vacuum pump to aid filtration. Figure 3 displays the filtration process.
7. In order to completely remove the HCl residues from the sample, gradually add 400 mL of deionized water to the filter paper with the sample and continue filtering.
8. Remove the sample from the filter paper and place it in a beaker (*see* **Note 9**).
9. Suspend the sample in 300 mL of deionized water (*see* **Note 9**).

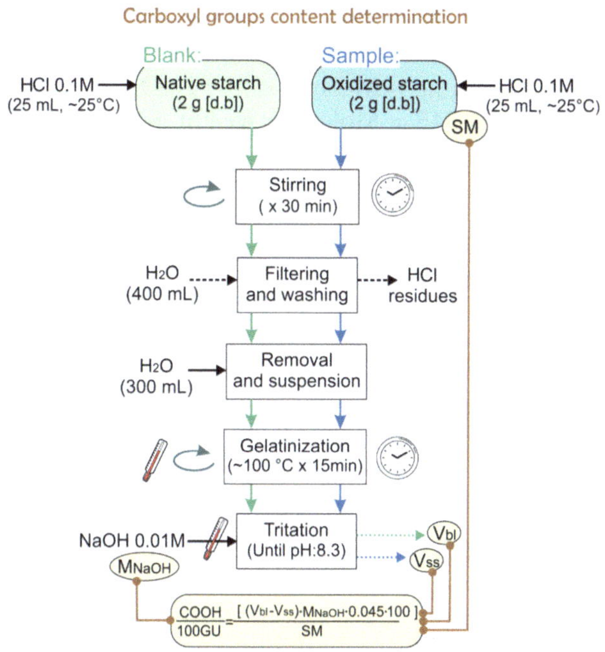

Fig. 4 Flowchart of the procedure to determine the content of carboxyl groups. Source: Authors' collection, 2025

10. Place the sample in a boiling water bath (~100°C) with constant stirring for 15 min (or until complete gelatinization of the starch samples) (*see* **Note 4**).
11. Also, carry out steps 4–10 in the blank.
12. Titrated the sample and the blank with the NaOH 0.01 M solution, until pH 8.3 (*see* **Notes 5, 6, and 10**). Record the total volume titrated to use this value for calculations of carboxyl group content.
13. In order to calculate the carboxyl group content, use the following equation:

$$\frac{\text{COOH}}{100\ \text{GU}} = \frac{[(V_{ss} - V_{bl}) \cdot M_{NaOH} \cdot 0.045 \cdot 100]}{\text{SM}}$$

Where

V_{ss} = Titration volume (mL) of the NaOH 0.01 M solution required for the oxidized samples;

V_{bl} = Titration volume (mL) of the NaOH 0.01 M solution required for the blank;
M_{NaOH} = Molarity of the NaOH solution (mol/L);
0.045 = Molecular weight of carboxyl unit (M(C) + 2 M(O) + M(H) = 0.045 kg/mol);
SM = Sample mass (g) on dry basis.

14. The result is expressed in the quantity of carboxyl groups per 100 glucose units (COOH/100 GU).
15. Figure 4 displays the flowchart of this analysis.

4 Notes

1. It is recommended to weigh the sample as reported in the description of the methodologies. However, if it is extremely necessary to reduce the amount of sample to be weighed, perform pretests to verify the reduction will not impair the quantification of carbonyl/carboxyl groups content.
2. In order to calculate the sample mass on a wet basis to obtain the specific amount on a dry basis, use the following equation:

$$M_{\text{dry basis}} = M_{\text{wet basis}} - \left[(M_{\text{wet basis}}) \times \frac{\text{MC}}{100} \right]$$

Where
$M_{\text{dry basis}}$ = sample mass (g) on dry basis;
$M_{\text{wet basis}}$ = sample mass (g) on wet basis;
MC = moisture content (%) of starch sample.

3. The analysis of carbonyl group content does not use native starch for the blank sample, once the native starch can present this compound in its composition. On the other side, the blank of analysis of carboxyl group content uses native starch, since it does not present the carboxyl group in its composition.
4. During gelatinization of starch, it is important to keep them under vigorous and constant stirring to ensure homogenization of the samples and avoid errors in the quantification. Starch granules are insoluble in water and, then, must be kept in suspension until complete gelatinization.
5. To control pH, use a calibrated potentiometer. In addition, samples must be in constant stirring during the pH adjustment or while titrating samples to ensure homogenization.

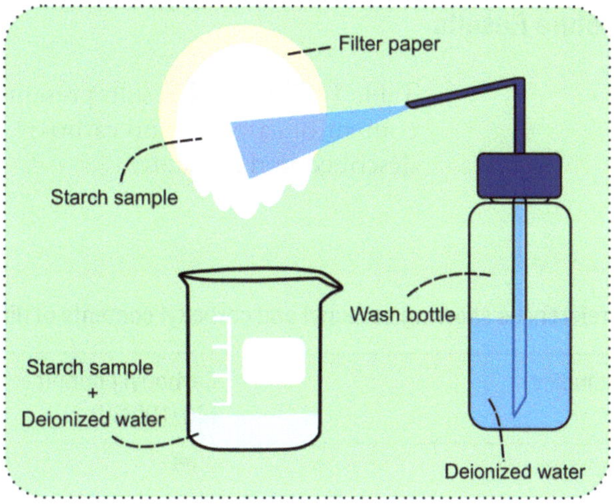

Fig. 5 Removal of the starch from the paper filter. Source: Authors' collection, 2025

6. The titration can be performed by using a buret (less expensive, but more laborious and less precise) or an automatic titrator (more expensive, but less laborious and more precise).
7. It is important to filter the samples immediately, since if there is a delay in performing step 5, the HCl solution continues to react with the starch, and it can cause errors in the quantification of the carboxyl content. Therefore, if there is only one vacuum pump, do not prepare several samples, but prepare one sample at a time.
8. Wet the filter paper with deionized water to facilitate its placement in the Büchner funnel, as well as to facilitate the filtration process.
9. Carefully use a small spatula to remove the starch from the filter paper. If there is still starch left on the filter paper, proceed as follows: place the deionized water (300 mL) that will be used to gelatinize the sample in a wash bottle. Then, use the wash bottle to remove the rest of the starch. It is important that the water used to remove the starch be placed in a beaker along with the starch sample. If there is water left in the wash bottle, place the remaining water inside the beaker together with the sample and then proceed with step 9. Figure 5 illustrates how to remove the starch from the filter paper.
10. It is recommended to titrate the samples while they are still hot, in order to avoid analysis errors. This is a very important step, and, in general, an automatic titrator is better suited to attain it than using a buret.

5 Literature Results

Table 1 shows some results presented in the literature regarding the content of carbonyl and carboxyl, considering the methodologies described in this chapter.

Table 1
Selected references showing carbonyl and carboxyl contents of different starch sources

Starch samples	Carbonyl content (CO/100 GU)	Carboxyl content (COOH/100 GU)	References
Native potato starch	0.04	–	[2]
Native arracacha starch	0.08	–	[19]
Potato starch treated with sodium hypochlorite (2%)	0.07	0.39	[5]
Corn starch treated with sodium hypochlorite (2%)	0.06	0.14	[20]
Rice starch treated with sodium hypochlorite (2%)	0.06	0.24	
Maize starch treated with sodium hypochlorite (2%)	0.38	0.19	[16]
Potato starch treated with ozone (30 min)	0.10	0.04	[2]
Cassava starch treated with plasma (30 kHz)	1.5	0.65	[21]
Cassava starch treated with ozone (30 min)	0.13	0.05	[22]
Cassava starch treated with dry heating (4 h)	0.22	0.00	[23]
Cassava starch treated with ultrasound (8 h, 25 kHz, 72 W/L)	0.15	0.01	[24]
Arracacha starch treated with ozone (30 min)	0.13	0.03	[25]
Carioca bean starch treated with heat-moist (22% of moisture content, 1 h, 100°C)	0.06	–	[26]
Carioca bean starch treated with sodium hypochlorite (1.5%)	0.09	0.11	[24]

Acknowledgments

The authors are grateful to São Paulo Research Foundation (FAPESP, Brazil) for funding the project n° 2019/05043-6 and the BC Maniglia Young Investigator grant (2021/05947-2); the "Coordenação de Aperfeiçoamento de Pessoal de Nível Superior—Brasil (CAPES, Brazil)"—Finance Code 8888.378469/2019-1, for financing the DC Lima Ph.D. scholarship. Communauté urbaine du Grand Reims, Département de la Marne, Région Grand Est, and European Union (FEDER Champagne-Ardenne 2014-2020, FEDER Grand Est 2021-2027) are acknowledged for their financial support to the Chair of Biotechnology of CentraleSupélec and the Centre Européen de Biotechnologie et de Bioéconomie (CEBB).

References

1. Nawaz H, Waheed R, Nawaz M, Shahwar D (2020) Physical and chemical modifications in starch structure and reactivity. In: Chemical properties of starch. IntechOpen. https://doi.org/10.5772/intechopen.88870
2. Castanha N, Matta Junior MD da, Augusto PED (2017) Potato starch modification using the ozone technology. Food Hydrocoll 66:343–356
3. Salmi T, Tolvanen P, Wärnå J, Mäki-Arvela P, Murzin D, Sorokin A (2016) Mathematical modeling of starch oxidation by hydrogen peroxide in the presence of an iron catalyst complex. Chem Eng Sci 146:19–25
4. Li J, Zhou M, Cheng F, Lin Y, Shi L, Zhu P-X (2020) Preparation of oxidized corn starch with high degree of oxidation by fenton-like oxidation assisted with ball milling. Mater Today Commun 22:100793
5. Kuakpetoon D, Wang Y (2001) Characterization of different starches oxidized by hypochlorite. Starch 53:211–218
6. Wurzburg OB (1986). Modified starches-properties and uses. CRC Press Inc, Boca Raton, FL, p 277
7. Isbell HS, Frush HL (1987) Mechanisms for hydroperoxide degradation of disaccharides and related compounds. Carbohydr Res 161:181–193
8. Kuakpetoon D, Wang Y-J (2008) Locations of hypochlorite oxidation in corn starches varying in amylose content. Carbohydr Res 343:90–100
9. Butrim SM, Bil'dyukevich TD, Yurkshtovich TL (2001) Kinetics of starch oxidation in the system nitrogen(IV) oxide-tetrachloromethane. Russ J Appl Chem 74:2106–2110
10. Barbosa JV, Martins J, Carvalho L, Bastos MMSM, Magalhães FD (2019) Effect of peroxide oxidation on the expansion of potato starch foam. Ind Crops Prod 137:428–435
11. Zhang Y-R, Wang X-L, Zhao G-M, Wang Y-Z (2012) Preparation and properties of oxidized starch with high degree of oxidation. Carbohydr Polym 87:2554–2562
12. Lewicka K, Siemion P, Kurcok P (2015) Chemical modifications of starch: microwave effect. Int J Polym Sci 2015:1–10
13. Raghunathan R, Pandiselvam R, Kothakota A, Mousavi Khaneghah A (2021) The application of emerging non-thermal technologies for the modification of cereal starches. LWT 138:110795
14. Maniglia BC, Castanha N, Le-Bail P, Le-Bail A, Augusto PED (2020) Starch modification through environmentally friendly alternatives: a review. Crit Rev Food Sci Nutr 0:1–24
15. Maniglia BC, Castanha N, Rojas ML, Augusto PED (2021) Emerging technologies to enhance starch performance. Curr Opin Food Sci 37:26–36
16. Dao PH, Nam TT, Phuc M Van, Hiep NA, Thanh T Van, Vuong NT, Xuan DD (2017) Oxidized maize starch: characterization and effect of it on the biodegradable films. II. Infrared spectroscopy, solubility of oxidized starch and starch film solubility. Vietnam J Sci Technol 55:395
17. Smith RJ (1967) Characterization and analysis of starches. In: Whistler RL, Paschall EF (eds). Starch: chemistry and technology, 1st edn. Academic Press, New York, pp 569–635
18. Chattopadhyay S, Singhal RS, Kulkarni PR (1997) Optimisation of conditions of synthesis

of oxidised starch from corn and amaranth for use in film-forming applications. Carbohydr Polym 34:203–212
19. Castanha N, Villar J, Matta Junior MD, Anjos CBPdos, Augusto PED (2018) Structure and properties of starches from Arracacha (Arracacia xanthorrhiza) roots. Int J Biol Macromol 117:1029–1038
20. Maniglia BC, Lima DC, Matta Júnior Mda, Oge A, Le-Bail P, Augusto PED, Le-Bail A (2020) Dry heating treatment: a potential tool to improve the wheat starch properties for 3D food printing application. Food Res Int 137:109731
21. Klanarong N (2020) Characterization of native and oxidized cassava starch prepared using a solution plasma process. Naresuan Univ Eng J 15:81–87
22. Lima DC, Castanha N, Maniglia BC, Matta Junior MD, La Fuente CIA, Augusto PED (2021) Ozone processing of cassava starch. Ozone Sci Eng 43:60–77
23. Maniglia BC, Lima DC, Matta Junior MD, Le-Bail P, Le-Bail A, Augusto PED (2020) Preparation of cassava starch hydrogels for application in 3D printing using dry heating treatment (DHT): a prospective study on the effects of DHT and gelatinization conditions. Food Res Int 128:108803
24. Castanha N, Lima DC, Matta Junior MD, Campanella OH, Augusto PED (2019) Combining ozone and ultrasound technologies to modify maize starch. Int J Biol Macromol 139:63–74
25. Lima DC, Villar J, Castanha N, Maniglia BC, Matta Junior MD, Augusto PED (2020) Ozone modification of arracacha starch: effect on structure and functional properties. Food Hydrocoll 108:106066
26. Evangelho JA, Biduski B, Silva WMF, El Halal SL, Lenhani GC, Pinto V, Dias ARG, Zavareze E (2021) Carioca bean starch upon synergic modification: characteristics and films properties. J Sci Food Agric 101:253–261

Chapter 12

Starch Paste Clarity Determination

Olaide A. Akintayo, Olayemi E. Dudu and Samson A. Oyeyinka

Abstract

Paste clarity is an important quality attribute of starch that significantly affects its suitability for diverse food and industrial applications, including beverages, sauces, and gels. It is largely determined by the interaction between amylose and water molecules under hydrothermal conditions, which influences starch gelatinization and the resulting paste's transparency or turbidity. Reliable assessment of paste clarity is critical for product formulation, quality control, and research into starch functionality. This chapter provides a guide and an updated protocol for measuring starch paste clarity using the spectrophotometric method. The method described offers a simplified and standardized approach that addresses variability in existing protocols, enabling consistent and reproducible measurements across different starch types and laboratories. Adoption of this protocol can enhance comparability of results, improve product development strategies, and support scientific investigations into starch physicochemical properties.

Keywords Amylose, Physicochemical properties, Spectrophotometry, Starch paste clarity

1 Introduction

Starch is an important ingredient with diverse applications in both food and non-food industries. Its utilization is influenced by its physicochemical properties, which depend largely on starch composition and structure. An important physicochemical property influencing the functionality of starch in food applications is paste clarity. Paste clarity refers to the ability of a starch paste to transmit light at a specified wavelength. A higher paste clarity indicates greater light transmission, resulting in a more transparent paste. The clarity of starch paste plays a key role in determining its suitability for specific products. For instance, starches that form opaque pastes are well-suited for applications like salad dressings, where visual opacity is often desirable. In contrast, starches that yield transparent pastes are ideal for products such as fruit pie fillings, where clarity enhances visual appeal [1–4].

Maria Teresa Pedrosa Silva Clerici and Pedro H. Campelo (eds.), *Basic Protocols in Starch Characterization*, Methods and Protocols in Food Science, https://doi.org/10.1007/978-1-0716-5116-2_12,

Paste clarity is affected by starch composition, particularly the amylose-to-amylopectin ratio, as well as minor components of starch such as lipids and proteins [5, 6]. High amylose starches tend to leach amylose during gelatinization into the surrounding medium, resulting in a cloudy, less clear paste, whereas starches rich in amylopectin generally produce clearer pastes [7]. Since native starches often lack desirable functional properties, such as resistance to shear and temperature, they are generally modified. This modification offers enhanced applications of starches and may change their physicochemical properties, including paste clarity. Notably, starch paste clarity varies greatly with botanical source [8, 9], starch modification method [10, 11] and wavelengths used in its determination [3, 12, 13], resulting in variability in reported data. The objective of this chapter is to promote a uniform, standardized protocol to improve consistency and comparison of starch paste clarity measurements worldwide.

The ability of starch components, amylose and amylopectin, to interact with light in the UV–visible region of the electromagnetic spectrum has been used to determine their presence in starch using spectrophotometric methods. Starch granules behave differently when cooked, and the resulting pastes show varying levels of light transmittance depending on the degree of association of starch chains [3]. Starch paste clarity measurement is simply governed by Beer–Lambert's law, which states that the amount of light absorbed by the medium (solution or sample) is directly proportional to the concentration (c) of the solute and the path length of radiation (l). When radiation falls on a homogenous medium, for example, a solution, a portion of the incident light is reflected (I_r), some portion is absorbed (I_a) by the medium, while the remaining is transmitted (I_t). Mathematically, the incident radiation (I_o) can be represented by Eq. (1):

$$I_o = I_r + I_a + I_t \tag{1}$$

where I_o = intensity of incident radiation, I_r = intensity of reflected radiation, I_a = intensity of absorbed radiation, and I_t = intensity of transmitted radiation.

If the radiation is passing through a medium made of glass, a negligible amount is reflected (I_r), thus the I_r in Eq. (1) can be eliminated. The equation can then be modified to give Eq. (2) as shown below. Figure 1 illustrates the passage of radiation through a medium or a solution with a negligible amount of reflected light.

$$I_o = I_a + I_t \tag{2}$$

where I_o = intensity of incident radiation, I_b = path length of radiation, and I_t = intensity of transmitted radiation.

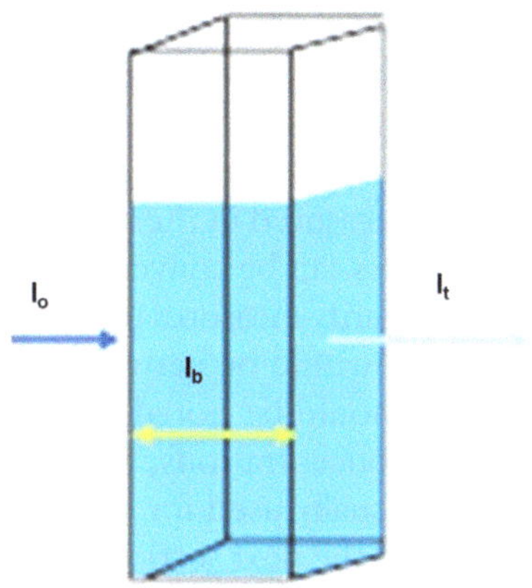

Fig. 1 Absorption of radiation through a medium or a solution

The Beer–Lambert's law can be expressed using Eq. (3) of a straight-line graph ($\Upsilon = mx$) and can be represented as shown below.

$$A = Kcl \tag{3}$$

where A is the absorbance of radiation, which is a measure of the amount of radiation absorbed by a medium, usually a solution, c is the concentration of the medium measured in mold m^{-3} and l is the path length of the solution, which is the distance traveled by the radiation through the solution measured in cm. The constant K is the Molar absorptivity, which is the absorbance measured for a solution having a concentration of 1 mold m^{-3} in a path length of 1 cm.

During starch paste clarity measurement, ultraviolet or visible light is passed through the starch solution which has been previously gelatinized. The paste clarity can then be measured in the transmittance or absorbance mode on a UV- or Visible Spectrophotometer. It is important to emphasize that absorbance and transmittance are related, therefore the equations for absorbance (4) and transmittance (5) can be substituted to give another Eq. (6).

$$A = \mathrm{Log}_{10}\left(I_o/I_t\right) \tag{4}$$

$$T = (I_t/I_o) \tag{5}$$

$$T = 10^{-A} \tag{6}$$

where A = absorbance of radiation, I_o = intensity of incident radiation, I_t = intensity of transmitted radiation and T = transmittance of radiation.

From the equations above, it means that if the paste clarity is measured in the absorbance mode on a spectrophotometer, the values can be converted to transmittance and vice versa. Because paste clarity measures the amount of light passing through the starch paste, it would be better to take measurements in transmittance mode. It should be noted that Eq. (3) only holds over a wide range of low concentrations, and this explains why a 1% starch concentration is usually used for starch paste clarity measurement, as will be shown in the next section.

2 Material

The following materials and equipment are required to assess starch paste clarity using a standard spectrophotometric method.

(i) Test sample, which may be native or modified starch (e.g., corn, cassava, potato, sago or rice), food grade or analytical-grade.

(ii) Distilled or deionized water for preparing starch dispersions.

(iii) Standard laboratory glassware, including beakers and measuring cylinders.

(iv) Heating apparatus such as a water bath (preferably a shaking water bath) or a hot plate with stirring capability, capable of maintaining temperatures up to 100°C.

(v) UV–visible spectrophotometer capable of measuring absorbance at 650 nm.

(vi) Compatible cuvettes, typically with a 1 cm path length.

3 Method

This section outlines a standardized protocol for preparing starch pastes and assessing their clarity using spectrophotometric measurement of transmittance at 650 nm. The procedure is designed to be broadly applicable to a variety of native and modified starches and provides reproducible conditions for gelatinization, cooling, and clarity assessment. Key precautions for instrumentation and sample handling are also included to ensure accurate and consistent results.

3.1 Starch Preparation and Heating Procedure

(i) Accurately prepare a 1% (w/v) starch suspension by weighing 1 g of starch into a 100 mL screw-capped centrifuge tube and making up the volume to the mark with distilled water. Alternatively, 0.5 g of starch can be weighed into a 50 mL tube and made up to the mark with distilled water.

(ii) Vortex-mix the starch suspension thoroughly for at least 5 min to ensure proper dispersion.

(iii) Place the tube in a shaking water bath preheated to 95°C and incubate for 30 min. If a shaking water bath is unavailable, maintain the starch suspension in a stationary water bath at 95°C and vortex-mix the tube at 5-min intervals throughout the incubation.

(iv) After heating, cool the starch paste rapidly to room temperature (~25°C) within 5 min by immersing the tube in cold water.

3.2 Measurement of Starch Paste Clarity (Transmittance)

(i) Turn on the UV–visible spectrophotometer at least 20 min prior to measurement to allow the lamp to warm up and stabilize.

(ii) Clean the cuvette surfaces with a lint-free cloth.

(iii) Set the spectrophotometer to transmittance mode (%*T*) and select a wavelength of 650 nm.

(iv) Fill a cuvette approximately three-quarters full with distilled water to serve as the blank.

(v) Insert the blank cuvette into the sample chamber and press the "zero" button to calibrate.

(vi) Remove the blank and fill a clean cuvette with the starch paste sample, ensuring it has the same path length.

(vii) Wipe the outside of the sample cuvette with a lint-free cloth, insert it into the sample chamber, and close the chamber door.

(viii) Record the transmittance (%*T*) value displayed.

4 Notes

Starches exhibiting different paste clarities have diverse food and industrial applications. The selection of wavelength significantly influences the extent of light transmission through a starch medium, which serves as a fundamental parameter in assessing starch paste clarity. Consequently, measurements obtained at different wavelengths may yield varying results, even when starch type and concentration are not changed. To avoid misleading conclusions or inappropriate starch selection based on reported paste clarity, global uniformity in spectrophotometric measurement, especially regarding wavelength selection, is essential. The proposed standardization proposed in this book will enhance consistency, reproducibility, and reliability of experimental results, facilitating comparability across different regions and facilities. It is anticipated that research laboratories and industries will soon adopt a standardized protocol for starch paste clarity analysis, and such a standardized procedure would be regulated by relevant bodies or agencies. Accurate measurements also depend on strict adherence to instrumental and procedural best practices. Therefore, the following precautions are recommended to ensure reliable and reproducible results, while maintaining the integrity of both samples and spectrophotometric equipment.

(i) Ensure the spectrophotometer is properly calibrated and maintained.

(ii) Always allow the instrument and lamp to warm up for at least 20 min before measurements.

(iii) Position cuvettes so the light beam passes through the clear walls.

(iv) Close the cell chamber door fully before taking readings.

(v) Clean any spills inside the cuvette compartment promptly.

(vi) Verify that the spectrophotometer is set to measure transmittance (%T) before recording data.

References

1. Akhila PP, Sunooj KV, Aaliya B, Navaf M, Sudheesh C, Yadav DN, Khan MA, Mir SA, George J (2022) Morphological, physicochemical, functional, pasting, thermal properties and digestibility of hausa potato (*Plectranthus rotundifolius*) flour and starch. Appl Food Res 2(2):100193
2. Bello-Pérez LA, Ortíz-Maldonado F, Villagómez-Mendez J, Toro-Vazquez JF (1998) Effect of fatty acids on clarity of starch pastes. Starch 50(9):383–386
3. Craig SS, Maningat CC, Seib PA, Hoseney R (1989) Starch paste clarity. Cereal Chem 66(3):173–182
4. Subroto E, Cahyana Y, Indiarto R, Rahmah TA (2023) Modification of starches and flours by acetylation and its dual modifications: a review of impact on physicochemical properties and their applications. Polymers 15(14):2990
5. Boonkor P, Sagis LM, Lumdubwong N (2022) Pasting and rheological properties of starch paste/gels in a sugar-acid system. Foods 11(24):4060
6. Pycia K, Gałkowska D, Juszczak L, Fortuna T, Witczak T (2015) Physicochemical, thermal and rheological properties of starches isolated from malting barley varieties. J Food Sci Technol 52(8):4797–4807
7. Oyeyinka SA, Singh S, Adebola PO, Gerrano AS, Amonsou EO (2015) Physicochemical properties of starches with variable amylose contents extracted from bambara groundnut genotypes. Carbohydr Polym 133:171–178
8. Kumari B, Sit N (2023) Comprehensive review on single and dual modification of starch: methods, properties and applications. Int J Biol Macromol 253:126952
9. Waterschoot J, Gomand SV, Fierens E, Delcour JA (2015) Production, structure, physicochemical and functional properties of maize,

cassava, wheat, potato and rice starches. Starch 67(1–2):14–29
10. Punia S (2020) Barley starch modifications: physical, chemical and enzymatic—a review. Int J Biol Macromol 144:578–585
11. Shen H, Yan M, Liu Y, Liu X, Ge X, Muratkhan M, Ospankulova G, Zhang G, Li W (2023) Multiscale structure-property relationships of oxidized wheat starch prepared assisted with electron beam irradiation. Int J Biol Macromol 235:123908
12. Hoover R, Hadziyev DJSS (1981) Characterization of potato starch and its monoglyceride complexes. Starch 33(9):290–300
13. Mellies R, Yeates T, Mehltretter C, Senti F (1961) Preparation and physical characteristics of cross-linked hypochlorite-oxidized corn starch. Starch 13(4):114–116

Chapter 13

Starch Syneresis

Gabriela Noel Barrera, Pablo Martin Palavecino and Pablo Daniel Ribotta

Abstract

The term syneresis describes the phenomenon whereby a polymeric matrix expels liquid from itself. Syneresis is a complex molecular process where an increase in the number of macromolecular junctions leads to network rearrangement, causing the interstitial liquid release. In starch systems, syneresis is specifically the phenomenon associated with water expelling out from gelatinized starch systems during cooling, storage, and freeze–thawing as a consequence of the starch retrogradation. The water phase separation from the retrograded starch polymeric–matrix is associated with food quality issues; therefore, the syneresis level is used as a stability indicator of starchy foods. Mainly, two procedures are used to quantify the released liquid from starch systems: centrifugation and vacuum methods, although the experimental procedures in both cases have not been standardized yet. Both methodologies use an external force to expel the water from the polymeric network, and the weight of the removed water is used as a syneresis measure. This chapter provides a general overview of syneresis in starch systems by presenting the main aspects of this phenomenon. The two main analytical methods to determine the degree of syneresis are presented, and a protocol for the centrifugation method and the vacuum method is proposed, taking several measurement conditions into account. Additionally, some recommendations are given for controlling the measurement conditions in order to prevent imprecise estimates of syneresis.

Keywords Starch, Syneresis, Retrogradation, Water release, Centrifugation, Vacuum filtration

1 Introduction

Syneresis is the phenomenon of water release from a polymeric matrix during storage, mainly due to molecule rearrangement. Particularly, starch cooking in water (gelatinization) and posterior cooling creates a carbohydrate tridimensional network (gel) which encloses the liquid. After these processes, during the subsequent cold storage, starch molecules reorganize their structure (retrogradation), shrinking the interstitial spaces and consequently leaching out part of the water [1, 2].

Syneresis level depends on a number of factors, such as the amylose/amylopectin ratio and length, starch concentration, pH, presence of electrolytes, and sugars and the cooking process and temperature of cooling storage [1, 3, 4].

Maria Teresa Pedrosa Silva Clerici and Pedro H. Campelo (eds.), *Basic Protocols in Starch Characterization*, Methods and Protocols in Food Science, https://doi.org/10.1007/978-1-0716-5116-2_13,

Several food products use native and modified starch as cheap and reliable thickeners and gelling agents, but syneresis could be a drawback [3]. This process affects the starch gels and represents a negative aspect of food, especially in food products where syneresis changes sensory attributes like texture and visual aspect. The expelled liquid contains the same solutes as those inside food systems, such as salts, sugars, and soluble polymers [2, 5]. Then, syneresis not only produces technological and sensory properties detriment, but also creates an environment susceptible to microorganism growth. Consequently, syneresis measurement of starch gels and starchy formulations is essential to product success.

In retrograded starch systems, the extent of syneresis is quantified from the exuded fluid. Even though the spontaneous water expulsion from starch systems is not visibly and easily detected as a liquid layer on top of the product, there are methods and strategies by which it is possible to obtain quantitative data as well as to accelerate and promote the phase separation process between solvent and starch polymer chains. According to this, the weight of the released liquid during storage or after freeze/thaw treatments of the starch matrices is directly a measure of its syneresis degree [2, 5]. To quantify the water released from starch gel, the excluded fluid is separated and removed from the solid phase by the application of an external force, and the syneresis degree is calculated from the weight of removed water and is expressed as a percentage of gel weight.

Syneresis degree depends directly on the starch retrogradation level, which in turn is influenced by the processing conditions of the starch gel production. The strategies aimed to favor the increase in the molecular order of gelatinized systems, making the spontaneous syneresis more easily detectable during the storage [5]. Maximizing the swelling degree of granules and the amylose leaching during starch cooking ensures a proper system retrogradation. Consequently, temperature (from gelatinization temperature to approximately 95°C) and time (enough to guarantee the complete starch gelatinization) during starch cooking are critical factors, as well as the shearing rate. On the other hand, the overcooking of the starch systems makes swollen granules more susceptible to mechanical disruption caused by shearing and could cause thermal hydrolysis, affecting the amylose and amylopectin reordering, and the water retention capacity of the system. In regard to the strategies aimed to favor retrogradation, starch polymers recrystallization could be enhanced by increasing starch concentration, slightly reducing the pH to 6.5, reducing the cooling process rate of the gelatinized samples, storing samples between 0°C and 5°C, and stressing systems under freeze–thaw cycles [4].

2 Material

2.1 Equipment

Precision scale (accurate to 0.001 g).

2.2 Centrifugation Method

Centrifuge and centrifuge tubes, preferably 15 mL tubes with a traditional plug seal cap.

2.3 Vacuum Filtration

Vacuum filtration system: vacuum pump, filter paper and vacuum flask (kitasato) coupled to a sintered glass funnel or an Erlenmeyer flask with a bottle-top vacuum filter.

2.4 Gel Preparation

Prepare starch pastes or gels using deionized water or solutions made with deionized water and analytical grade reagents.

Use a temperature-controlled water bath with adequate agitation, such as a shaking water bath or a stirring hot plate. The use of devices such as Rapid Viscosity Analyzer™ (RVA) or viscoamylograph are also valid (*see* **Note 1**). If the gelatinization is not performed in sealed tubes, then the gelatinized starch suspension (paste) has to be poured into a container that must be hermetically sealed (*see* **Note 2**). The heating time must guarantee complete gelatinization. Minimum time is 10 min, but can vary according to the type of starch, suspension quantity and production method (*see* **Note 3**).

The starch concentration of the paste should be between 5% and 10% (*see* **Notes 4 and 5**).

2.5 Storage Conditions

The samples should be immediately cooled to room temperature and placed into the selected cold storage device. A fridge at 4°C must be used in order to assess cold storage water retention with proper cold storage times of 1 to 14 days. A freezer at −20°C should be chosen to perform sample freeze–thaw stability assays, and storage time should assure the complete freezing of the sample for each cycle (from 20 min to 24 h depending on gel size and freezer) (*see* **Note 6**).

3 Method

3.1 Centrifugation Method

Centrifugation is used as an external force to help expel the exuded water from gelatinized starch systems, favoring the condition for liquid separation. The extent of water release from starch gel or pasta is influenced by the force applied during centrifugation because the higher the centrifugal force, the more water pushes out from the polymer matrix. The centrifugation procedure is carried out under many different conditions, which include centrifugation force from

1500 to 8000× *g*, centrifugation time from 10 to 30 min and centrifugation temperature at 4, 20, or 25°C. Despite this, the centrifugation conditions often involve a centrifugation force from 2500 to 3500× *g* during 10 to 30 min at 25°C. The sample mass is another important variable; thus, it should be considered. In this regard, starch gel or paste portions from 10 to 30 g are usually centrifuged under diverse centrifugation conditions.

Considering numerous centrifugations procedures [6–20], the amount of water released from starch gel or pastes after centrifugation could be determined under the following protocol.

- Weigh the centrifuge tube with a stopper.
- Weigh 20 g gelatinized starch sample into the centrifuge tube.
- Centrifugate the sample, tempered at room temperature, at 3000× *g* for 20 min at 25°C (not including time to achieve speed). Allow the centrifuge to stop without braking.
- Carefully, discard the excluded water (supernatant liquid) using a tissue paper. If the sample is rigid enough (solid sediment), drain the centrifuge tube at a 45° angle for 5 min on tissue paper in the test tube rack. Blot again if necessary.
- Immediately stopper the tube, weigh the tube and gel.
- Calculate the weight difference of the gel before and after centrifugation (a measure of removed water).
- Calculate syneresis as a weight % of removed water based on the initial gel portion (*see* **Note 8**).

3.2 Vacuum Filtration

The vacuum filtration of gelatinized starch systems helps to expel the exuded water from them, facilitating the conditions for liquid separation. In order to accelerate the water release during the filtration, a constant weight (1 kg stainless steel cylinder) is usually placed upon the starch gel, which is usually put on top of the gel, taking care not to fracture the gel. The sensitivity of the vacuum method is influenced by the vacuum pressure and the time during which the vacuum pressure is applied. Low vacuum is used during the water filtration. Usually, the vacuum filtration conditions are carried out at room temperature in the pressure range from 3 to 92 kPa and filter time range from 5 to 10 min. Previously to the water suction, starch gel or paste is put on a filter paper sheet, which usually has a pore size from 11 to 25 μm (medium or fast flow filter-paper). The mass of the samples should be taken into account; starch gel or paste portions from 20 to 40 g are usually filtered under vacuum pressure.

Considering many vacuum filtration procedures [6, 21–26], the amount of water release of starch gel or pastes after vacuum filtration could be determined under the following protocol.

- Weigh 30 g gelatinized starch sample in a container (*see* **Note 7**).
- Weigh an empty glass funnel fitted with a sheet of filter paper (cellulose filter paper for gelatinous precipitates with 20–25 μm porous diameter).
- Carefully, transfer the sample tempered at room temperature to the glass funnel.
- Weigh the funnel plus a sheet of filter paper and gel.
- Put the glass funnel containing the sample on the vacuum filtration system.
- Vacuum-filter the samples at 85 kPa for 10 min to remove the water released by the gel.
- Immediately after filtration, weigh the funnel plus the sheet of filter paper and gel (*see* **Note 8**).
- Calculate the weight difference of the gel before and after filtration (a measure of water removed).
- Calculate syneresis as a weight % of removed water based on the initial gel portion (*see* **Note 9**).

4 Notes

1. Starch gelatinization in water could be achieved by several different procedures. We suggest that the analyst carry out a method representative of the process that the starch will undergo during its application. The temperature profile and the shear stress are key variables to successfully achieve this goal.
2. Low paste temperatures during the pouring step could lead to early gelification, mainly in highly concentrated suspensions, and consequently, deficient gel formation.
3. As a guide, 30 min is enough to gelatinize up to 100 mL of 10% w/w starch suspensions under stirring or shaking in a boiling water bath. In the RVA, the standard method is the most frequently used device profile to gelatinize the starch samples in order to test their syneresis.
4. Starch content is a key factor for the water gel behavior; consequently, to deeply understand the sample, several concentrations must be assessed.
5. Different salts (sodium benzoate, potassium sorbate, sodium azide) can be added at 0.1% to the paste sample to avoid microbial activity during repeated refrigeration or freeze–thaw cycles.
6. Starch retrogradation is influenced by storage time and temperature. Then, we recommend analyzing retrogradation kinetics

at different temperatures by differential scanning calorimetry (DSC) in order to aid in the interpretation of syneresis results. Also, it is recommended to perform a large number of freeze–thaw cycles to fully characterize starches.

7. The container should have a diameter of approximately 3 cm in order to produce an easy handle gel.

8. It is recommended not to remove the filter paper to take the gel weight after filtration because this procedure could cause the loss of part of the gel mass. Trying to remove the filter paper could worsen the analytical determination rather than improve it, leading to errors. Consequently, the water retained by the filter paper after the vacuum filtration of the sample could be considered not significant. Syneresis degree should be determined at least by triplication.

9. Despite the sample preparation and storage conditions providing favorable circumstances for liquid separation, the sensitivity of the method is often low, and the error associated with the values is often high.

References

1. Biliaderis CG (2009) Structural transitions and related physical properties of starch. In: Starch. Elsevier, pp 293–372
2. Eliasson A-C (2010) Gelatinization and retrogradation of starch in foods and its implications for food quality. In: Chemical deterioration and physical instability of food and beverages. Elsevier, pp 296–323
3. Mason WR (2009) Starch use in foods. In: Starch. Elsevier, pp 745–795
4. Ai Y, Jane J (2018) Understanding starch structure and functionality. In: Starch in food. Elsevier, pp 151–178
5. Mizrahi S (2010) Syneresis in food gels and its implications for food quality. In: Chemical deterioration and physical instability of food and beverages. Elsevier, pp 324–348
6. Zheng GH, Sosulki FW (1998) Determination of water separation from cooked starch and flour pastes after refrigeration and freeze-thaw. J Food Sci 63:134–139
7. Tovar J, Melito C, Herrera E, Rascón A, Pérez E (2002) Resistant starch formation does not parallel syneresis tendency in different starch gels. Food Chem 76:455–459
8. Lee H, Kim H-S (2020) Pasting and paste properties of waxy rice starch as affected by hydroxypropyl methylcellulose and its viscosity. Int J Biol Macromol 153:1202–1210
9. Shang J, Li L, Zhao B, Liu M, Zheng X (2020) Internationaljournal of biological macromolecules comparative studies on physicochemical properties of total, A- and B-type starch from soft and hard wheat varieties. Int J Biol Macromol 154:714–723
10. Liu S, Shen M, Xiao Y, Luo Y, Xie J (2020) Effect of maize, potato, and pea starches with Mesona chinensis polysaccharide on pasting, gelatinization properties, granular morphology and digestion. Food Hydrocoll https://doi.org/10.1016/j.foodhyd.2020.106047
11. Anbarani NM, Mohammad S, Razavi A, Taghizadeh M (2021) Food hydrocolloids impact of sage seed gum and whey protein concentrate on the functional properties and retrogradation behavior of native wheat starch gel. Food Hydrocoll 111:106261
12. Marboh V, Mahanta CL (2021) International journal of biological macromolecules physicochemical and rheological properties and *in vitro* digestibility of heat moisture treated and annealed starch of sohphlang (*Flemingia vestita*) tuber. Int J Biol Macromol 168:486–495
13. Deetae P, Shobsngob S, Varanyanond W, Chinachoti P, Naivikul O, Varavinit S (2008) Preparation, pasting properties and freeze–thaw stability of dual modified crosslink-phosphorylated rice starch. Carbohydr Polym 73:351–358

14. Charoenkul N, Uttapap D, Pathipanawat W, Takeda Y (2011) Physicochemical characteristics of starches and flours from cassava varieties having different cooked root textures. LWT Food Sci Technol 44:1774–1781
15. Srichuwong S, Isono N, Jiang H, Mishima T, Hisamatsu M (2012) Freeze-thaw stability of starches from different botanical sources: correlation with structural features. Carbohydr Polym 87:1275–1279
16. Yang L, Zhou Y, Wu Y, Meng X, Jiang Y, Zhang H, Wang H (2016) Preparation and physicochemical properties of three types of modified glutinous rice starches. Carbohydr Polym 137:305–313
17. Palavecino PM, Penci MC, Ribotta PD (2020) Effect of sustainable chemical modifications on pasting and gel properties of sorghum and cassava starch. Food Bioprocess Technol 13. https://doi.org/10.1007/s11947-019-02381-0
18. Li J, Zhou M, Cheng F, Lin Y, Hu J, Zhu P (2019) Characterization and properties of long-chain fatty acid starch esters prepared with regenerated starch by dry method. Starch 71:1900143
19. Zhang Y, Li B, Zhang Y, Xu F, Zhu K, Li S, Tan L, Wu G, Dong W (2019) Effect of degree of polymerization of amylopectin on the gelatinization properties of jackfruit seed starch. Food Chem 289:152–159
20. Liu Y, Gao J, Wu H, Gou M, Jing L, Zhao K, Zhang B, Zhang G, Li W (2019) Molecular, crystal and physicochemical properties of granular waxy corn starch after repeated freeze-thaw cycles at different freezing temperatures. Int J Biol Macromol 133:346–353
21. Takahashi S, Seib PA (1988) Paste and gel properties of prime corn and wheat starches with and without native lipids. Cereal Chem 65:474–483
22. Takahashi S, Ebihara M, Kainuma K (1998) Properties and gel characteristics of flour and starch obtained from newly developed rice cultivars. J Home Econ Jpn 49:235–241
23. Lee MH, Baek MH, Cha DS, Park HJ, Lim ST (2002) Freeze-thaw stabilization of sweet potato starch gel by polysaccharide gums. Food Hydrocoll 16:345–352
24. Takeiti C, Fakhouri F, Ormenese R, Steel C, Collares F (2007) Freeze-thaw stability of gels prepared from starches of non-conventional sources. Starch/Staerke 59:156–160
25. Pongjanta J, Utaipattanaceep A, Naivikul O, Piyachomkwan K (2009) Debranching enzyme concentration effected on physicochemical properties and a-amylase hydrolysis rate of resistant starch type III from amylose rice starch. Carbohydr Polym 78:5–9
26. Garrido LH, Schnitzler E, Zortéa MEB, de Souza Rocha T, Demiate IM (2014) Physicochemical properties of cassava starch oxidized by sodium hypochlorite. J Food Sci Technol 51:2640–2647

Check for updates

Chapter 14

Scanning Electron Microscopy in Starch Characterization

Elisa Cristina Andrade Neves, Roberto de Freitas Neves, Maria Teresa Pedrosa Silva Clerici and Rebeca Salvador-Reyes

Abstract

Scanning electron microscopy (SEM) is a robust technique to characterize the micro and nanostructure of starches and flours. This chapter presents a practical, step-by-step protocol, from specimen preparation to image capture and interpretation, to assess native granule morphology and processing-induced changes. Samples (powders or tissue fragments) are sectioned, dried below the starch gelatinization temperature (≈40–45°C) or lyophilized, stored over desiccant, mounted on carbon tape stubs, and sputter-coated with gold or carbon to improve surface conductivity. Imaging is carried out under low vacuum with progressive magnifications (e.g., 100× to >5000×), using secondary electrons for topographic contrast and backscattered electrons for atomic-number contrast; optional EDS can complement surface analysis. The protocol details focusing, lens alignment, and recording parameters, and highlights common artifacts (charging and residual moisture) and their prevention. Representative micrographs illustrate how SEM reveals granule size, shape, distribution, surface features, and growth rings, which are influenced by botanical origin and amylose/amylopectin ratio. The workflow also supports monitoring enzymatic, chemical, and physical modifications. Together, these procedures provide reliable, high-resolution images, and practical guidance for consistent morphological evaluation of starch and flour systems.

Keywords Scanning electron microscopy, Starch granule morphology, Sample preparation, Sputter coating

1 Introduction

Starch granules consist of semicrystalline assemblies whose amylose-to-amylopectin ratio varies with botanical origin and strongly influences their physicochemical, rheological, and morphological properties. While bulk and molecular attributes can be probed by thermal, rheological, and compositional analyses, direct assessment of granule morphology, and of processing-induced alterations, benefits from scanning electron microscopy (SEM). SEM employs a focused electron beam that raster-scans the specimen surface; electron–matter interactions generate multiple signals that are collected by dedicated

Maria Teresa Pedrosa Silva Clerici and Pedro H. Campelo (eds.), *Basic Protocols in Starch Characterization*, Methods and Protocols in Food Science, https://doi.org/10.1007/978-1-0716-5116-2_14,

detectors to produce high-resolution images and complementary compositional information [1, 2].

For imaging, secondary electrons (SE, <50 eV) provide topographic contrast that reveals surface relief, fracture features, and fines adhering to granules, whereas backscattered electrons (BSE) yield atomic-number contrast useful for distinguishing phases or deposits of different mean *Z*. When coupled with energy-dispersive X-ray spectroscopy (EDS), SEM can further interrogate elemental signatures on the granule surface. Owing to its large depth of field (on the order of ~10 μm at ~10,000× and up to ~1 cm at ~20×), SEM delivers crisp, well-focused micrographs across irregular surfaces, and the stage can be tilted and rotated to visualize three-dimensional aspects of granule architecture [3]. Compared with the ≈0.2 μm resolution limit of conventional optical microscopy, SEM enables nanoscale inspection of surface features [4].

SEM micrographs support detailed evaluation of granule size, size distribution, and shape—parameters that are strongly modulated by botanical source and, consequently, aid botanical identification [5, 6]. They also permit observation of growth rings and other structural motifs linked to starch biosynthesis and granule development stage [4, 7]. Beyond native morphology, SEM is a sensitive tool to monitor surface erosion, pitting, cracking, and aggregation arising from enzymatic, chemical, or physical modifications under diverse processing conditions [8].

This chapter consolidates a practical SEM workflow for starches and flours, from sample preparation (sectioning, low-temperature drying/lyophilization, mounting, and conductive coating) to image acquisition across progressive magnifications and data interpretation, highlighting common artifacts (e.g., charging and residual moisture) and best practices for reproducible, publication-quality micrographs.

2 Material

2.1 Material

- SEM stubs.
- Double-sided conductive carbon tape.
- Scalpel.
- Tweezers.
- Nitrile gloves.
- Air blower/compressed, oil-free air.

2.2 Equipment

- Drying oven.
- Desiccator with fresh silica gel.
- Gold sputter coater (Emitech, model K550X).

- Carbon coater (Quorum, model Q150T-ES).
- Scanning Electron Microscope (SEM).

2.3 General Information

Always consult the instrument manuals for operating conditions and safety. Ensure samples are completely dry before imaging to avoid charging, poor contrast, and potential instrument damage.

3 Method

3.1 Sample Preparation

1. Carefully section native tissues longitudinally or transversely or fractionate samples to expose both surface and interior.
2. Dry fragments or powders by lyophilization or in a forced-air oven below the gelatinization temperature of the starch (≈40–45°C) to prevent morphological changes.
3. Store dried samples in a desiccator until analysis. Note: Residual moisture causes image artifacts and may harm the equipment (cf. Fig. 1A1, A2 vs. Fig. 1B1, B2).

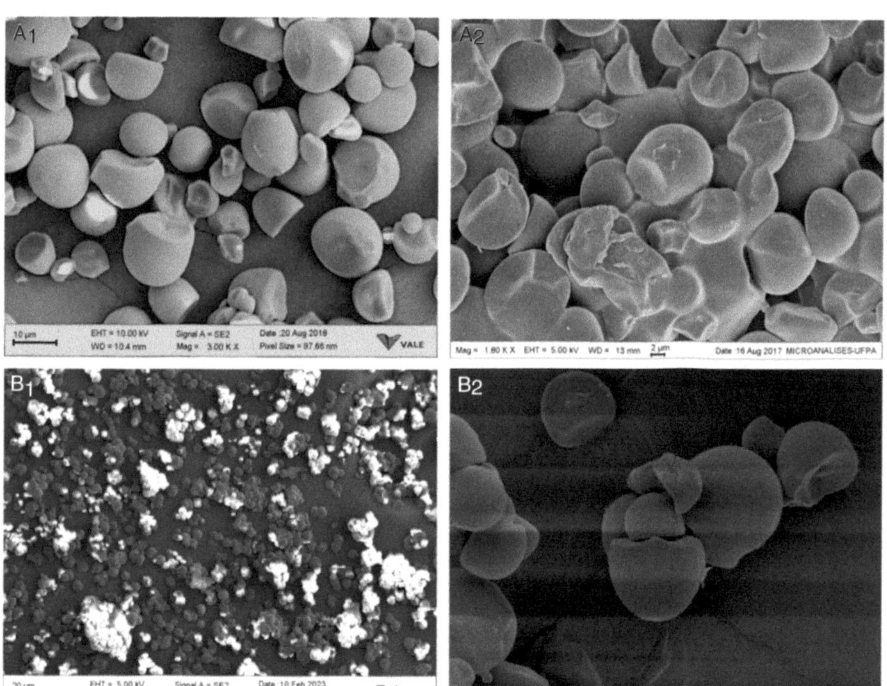

Fig. 1 Scanning electron micrographs (SEM) of starch samples: (**A1**, **A2**) well-resolved images from properly dried specimens; (**B1**, **B2**) degraded image quality due to residual moisture (charging, streaking, and loss of contrast). Source: Authors' collection, 2025

3.2 Stub Preparation

1. Label each stub and record the sample ID.
2. Affix a small piece of double-sided carbon tape to the stub surface.
3. Deposit a thin layer of dry powder and remove excess with an air blower.
4. For fragments, adhere small pieces to the tape. Orientation tip: Mount the flat face down to visualize external relief; mount the cut face up to observe internal features.

3.3 Sputter Coating (Metallization)

Purpose: Increase surface conductivity and secondary-electron emission.

1. Transfer mounted stubs to the *gold or carbon coater* (Fig. 2).
2. Typical conditions (or follow manufacturer instructions):
   - *Gold*: SE/BSE/EDS imaging, 1.5 min (~15 nm); cathodoluminescence, 30 s (~5 nm).
   - *Carbon*: SE/BSE/EDS—5 pulses × 2 s (~20 nm).
3. Image-coated samples may be stored immediately or in a desiccator (samples may be hygroscopic).

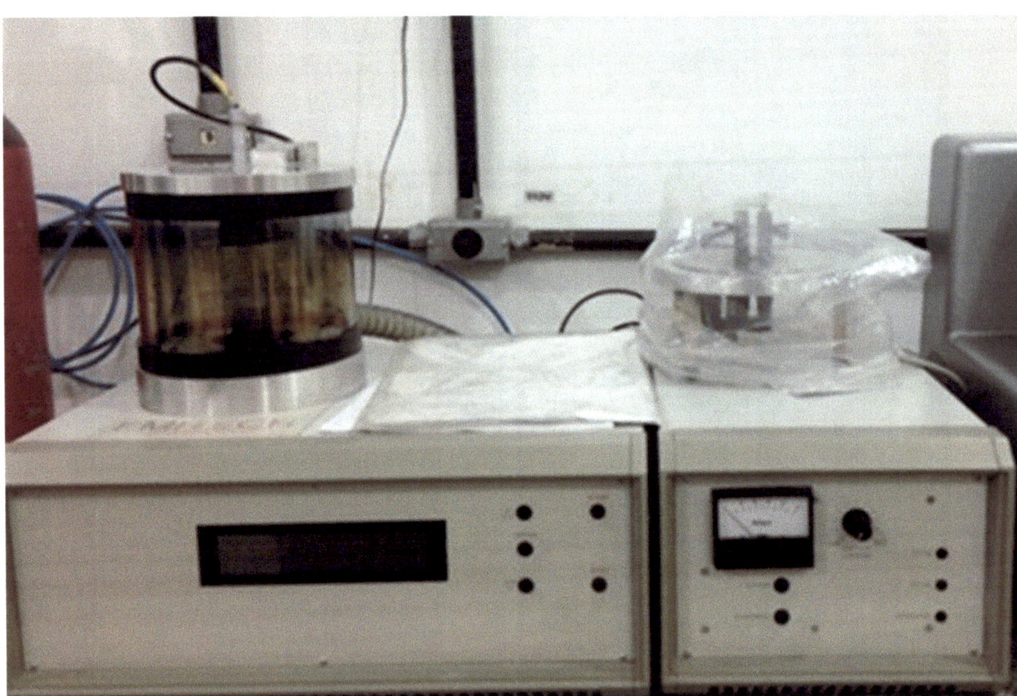

Fig. 2 Gold sputter coater (Emitech K550X) used to deposit a thin Au film (~15 nm) on mounted specimens. The vacuum chamber is shown on the left; the power/control unit with the stub holder is on the right. Source: Authors' collection, 2025

3.4 SEM Image Acquisition

The SEM comprises the electron column (source and lenses), a low-vacuum sample chamber (≈1–10 Torr) for the prepared stubs, and detectors.

1. Load stubs into the SEM carrousel (Fig. 3) and note each position.
2. For each sample, perform lens alignment, focus, and astigmatism correction; select detector (SE for topography, BSE for Z-contrast).
3. Start at low magnification (e.g., 100×), capture and save; increase stepwise (500×, 1500×, and >5000×) to resolve surface features (see Fig. 4).

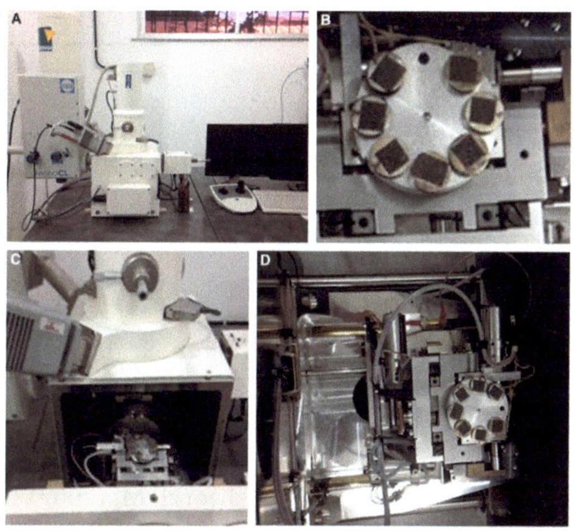

Fig. 3 Scanning electron microscope (SEM) and main components: (**A**) overall view of the instrument with the electron column and control station; (**B**) multistub carrousel; (**C**) sample chamber open, showing the stage/sample holder; (**D**) close-up of the stage in low-vacuum configuration with the carousel mounted. Source: Authors' collection, 2025

Fig. 4 SEM micrographs of the endosperm and starch granules of Andean maize Chullpi at three magnifications: 100× (overall endosperm view), 500× (granule distribution and emergence), and 1500× (granule morphology and surface contours). Typical conditions: SE detector, EHT 10 kV, WD 12–13 mm; scale bars as shown (100 μm at 100×; 20 μm at 500× and 1500×)

4. Record accelerating voltage (EHT), beam current/spot size, working distance (WD), detector, chamber pressure, and magnification in the image metadata.
5. At the end, document the SEM model and the final operating conditions.

3.5 Waste Management

Waste management must be carried out in accordance with **Note 1**.

3.6 Micrograph Analysis

Assess granule shape, size, and size distribution, surface relief/defects (pits, fissures, erosion), and features related to botanical origin (e.g., growth rings). Compare fields across magnifications and replicate stubs to confirm observations and minimize preparation artifacts.

4 Note

1. Dispose of carbon tape with residual sample in appropriate solid-waste containers according to institutional EHS guidelines.

References

1. Horovitz O, Cioica N, Jumate N, Pojar-Feneşan M, Balea A, Liteanu V, Mocanu A, Tomoaia-Cotişel M (2011) SEM characterization of starch granules. Stud UBB Chem 56(1)
2. Egerton RF (2005) The scanning electron microscope. In: Physical principles of electron microscopy: an introduction to TEM, SEM, and AEM. Springer US, Boston, MA, pp 125–153
3. DeCastro, LAS (2002). Sample processing for scanning electron microscopy. Embrapa Clima Temperado, Pelotas, Brazil, p 37
4. Horovitz O, Cioica N, Jumate N, Pojar-Feneşan M, Balea A, Liteanu V, Mocanu A, Tomoaia-Cotişel M (2011) SEM characterization of starch granules. Stud UBB Chem 56(1):211
5. Pérez S, Bertoft E (2010) The molecular structures of starch components and their contribution to the architecture of starch granules: A comprehensive review. Starch-Stärke 62(8):389–420
6. Leonel M (2007) Analysis of the shape and size of starch granules from different botanical sources. Food Sci Tech 27:579–588
7. Chakraborty I, Pallen S, Shetty Y, Roy N, Mazumder N (2020) Advanced microscopy techniques for revealing molecular structure. Biophys Rev 12:105–122
8. Yan H, Gu Z (2010) Morphology of modified starches prepared by different methods. Food Res Int 43(3):767–772

Check for updates

Chapter 15

Morphology and Size Distribution of Starch Granules: A Sem–Imagej Quantitative Analysis Workflow

Rebeca Salvador-Reyes, Elisa Cristina Andrade Neves, Roberto de Freitas Neves, and Maria Teresa Pedrosa Silva Clerici

Abstract

Quantitative analysis of SEM micrographs is essential to convert qualitative observations of starch granules into objective descriptors of size and shape. This chapter presents a reproducible SEM–ImageJ workflow that uses calibrated SEM images (JPG/TIF) to measure morphology and size distribution. The protocol covers scale calibration from the micrograph bar; preprocessing (8-bit conversion, contrast enhancement, light smoothing); segmentation (thresholding and watershed) to isolate touching granules; ROI curation; and measurement via Analyze Particles. Recommended outputs include area, perimeter, equivalent circular diameter (ECD), Feret max/min, aspect ratio, circularity, and solidity. We provide guidance on sampling (multiple fields; ≥200–300 granules), artifact control (edge objects, charging, compression), and standardized reporting of histograms, cumulative curves, and summary statistics. Metadata capture and macro saving are emphasized to ensure reproducibility and comparability across samples, instruments, and treatments.

Keywords SEM, ImageJ, Starch granules, Morphology, Size distribution, Feret diameter, Circularity

1 Introduction

Starch granules display wide variability in size, shape, and surface features that reflect botanical origin and the relative amounts and organization of amylose and amylopectin. These morphological attributes are closely linked to hydration, gelatinization, and processing performance, making them valuable targets for quantitative characterization [1–3]. Scanning electron microscopy (SEM) provides high-resolution, high-depth-of-field images that reveal granule architecture and surface relief, but qualitative description alone is insufficient for rigorous comparison across samples or treatments [4].

This chapter presents a SEM–ImageJ quantitative analysis workflow to convert micrographs into numerical descriptors of granule morphology and size distribution. Building on the previous chapter

Maria Teresa Pedrosa Silva Clerici and Pedro H. Campelo (eds.), *Basic Protocols in Starch Characterization*, Methods and Protocols in Food Science, https://doi.org/10.1007/978-1-0716-5116-2_15,

(SEM image acquisition), we detail a calibrated image-analysis pipeline consistent with particle image-analysis standards [4]: scale calibration from the micrograph's bar, field selection, preprocessing (noise reduction and contrast enhancement), segmentation (thresholding and watershed to separate touching granules), and measurement. The protocol yields primary geometric metrics—area, perimeter, equivalent circular diameter (ECD), Feret diameters (max/min), aspect ratio, roundness, and circularity—and prescribes best practices for reporting distributions (histograms and cumulative curves), summary statistics, and uncertainty.

It was adopting ImageJ/Fiji, an open-source platform with validated performance and broad community support, which enables reproducible analysis via macros and batch processing [5, 6]. Attention is given to sources of bias common in starch micrographs—overlapping particles, anisotropic illumination, charging artifacts, and 2D projection of 3D objects—and to strategies for robust sampling (analyzing multiple, well-dispersed fields and hundreds of particles when feasible) [5, 7, 8]. The result is a transparent, portable method to quantify starch–granule morphology and size distributions for research, quality control, and process optimization.

2 Material

2.1 Inputs and Software

2.1.1 Required Inputs (SEM Micrographs)

- SEM images of starch granules in JPEG (.jpg) as provided (acceptable), preferably also available in TIFF (.tif) to avoid compression artifacts.
- Image characteristics: 8-bit grayscale; resolution ≥1024 × 1024 px (≥2048 × 2048 px recommended); even illumination; in-focus; no saturation.
- Calibration: each image must include a scale bar or a known pixel size from the SEM metadata. Do not resample or crop out the scale bar before calibration.
- Magnifications: include low (~100×) for context, medium (500–1500×) for size distribution, and high (>5000×) for surface features (as available).
- Sampling: analyze multiple, nonoverlapping fields per sample and ≥200–300 granules per condition when feasible [5].

2.1.2 Software

- ImageJ (NIH ImageJ 1.x) or Fiji (recommended; includes plugins and Java).
- Version: ImageJ 1.53+/ Fiji (current stable).

2.1.3 Practical Considerations

- Prefer TIFF for archiving/analysis; if using JPEG, keep the highest quality to minimize compression blocks that hinder thresholding.
- Maintain original filenames and keep an unaltered copy of each micrograph.
- Record SEM metadata (magnification, working distance, detector, accelerating voltage) alongside each image.

3 Method

3.1 Procedure

Adapted from Oliveira et al. [9] and aligned with [5] (static image analysis).

3.2 Basic Steps

- **Open the SEM Micrograph**
  In ImageJ, go to File ▶ Open, and select the.jpg or.tif file as described in Fig. 1.
- *Calibrate the image scale* (critical for all measurements)
  - With the Straight Line tool, draw a line exactly over the micrograph's scale bar as shown in Fig. 2.
  - Go to Analyze ▶ Set Scale....

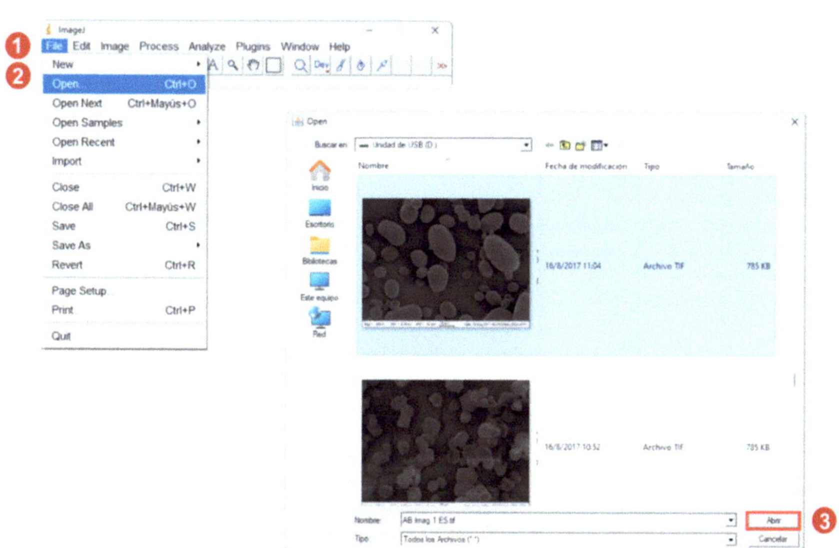

Fig. 1 Opening file in imageJ software. Source: Authors' collection, 2025

- In the dialog, enter the Known distance (e.g., 20) and the Unit of length (e.g., μm). Leave Pixel aspect ratio = 1.0.
- Check Global if you want the same scale applied to subsequent images in the session. Click OK.

  Tip: Do not crop out the scale bar before calibration.

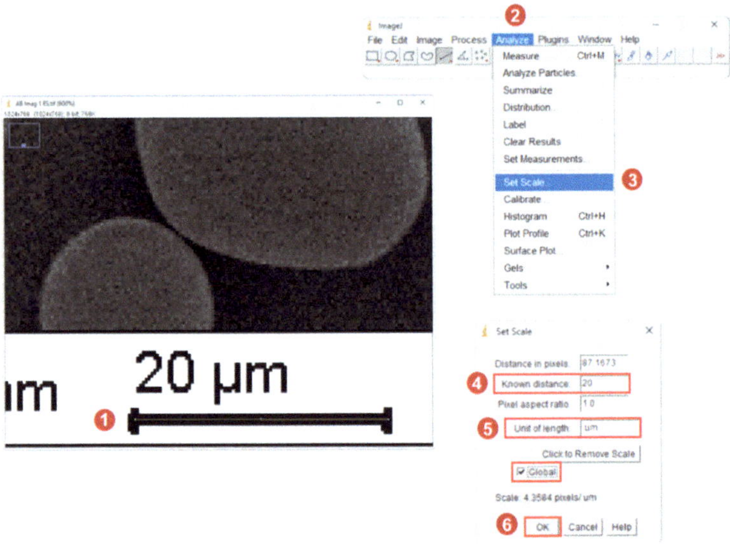

Fig. 2 Image scale calibration. Source: Authors' collection, 2025

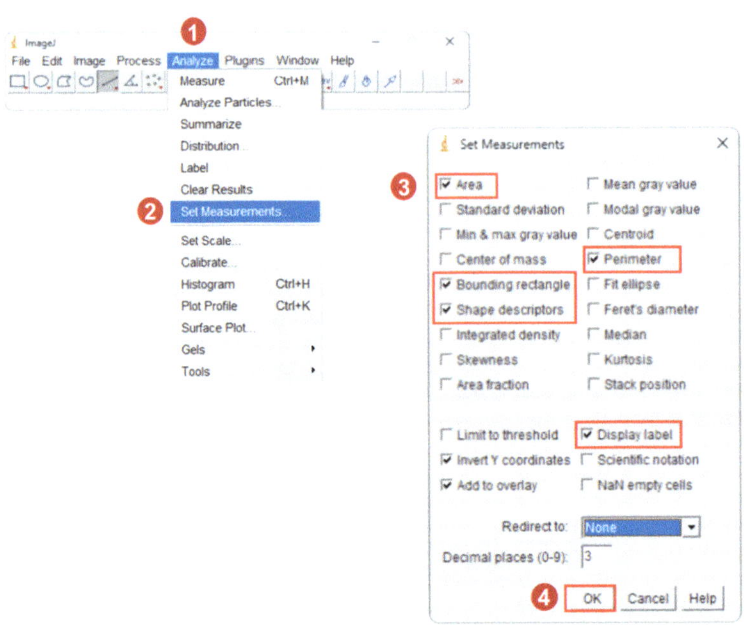

Fig. 3 Choose the measurement outputs. Source: Authors' collection, 2025

- *Choose the measurement outputs*

Go to Analyze ▶ Set Measurement, and select at least: area, perimeter, Feret's diameter, fit ellipse (major/minor), shape descriptors (circularity, aspect ratio, roundness), bounding rectangle, and display label. Set Decimal places as needed (e.g., 3). Click OK as shown in Fig. 3.

- *Prepare the image for accurate particle selection*
    - Convert to grayscale: Image ▶ Type ▶ 8-bit, as shown in Fig. 4.

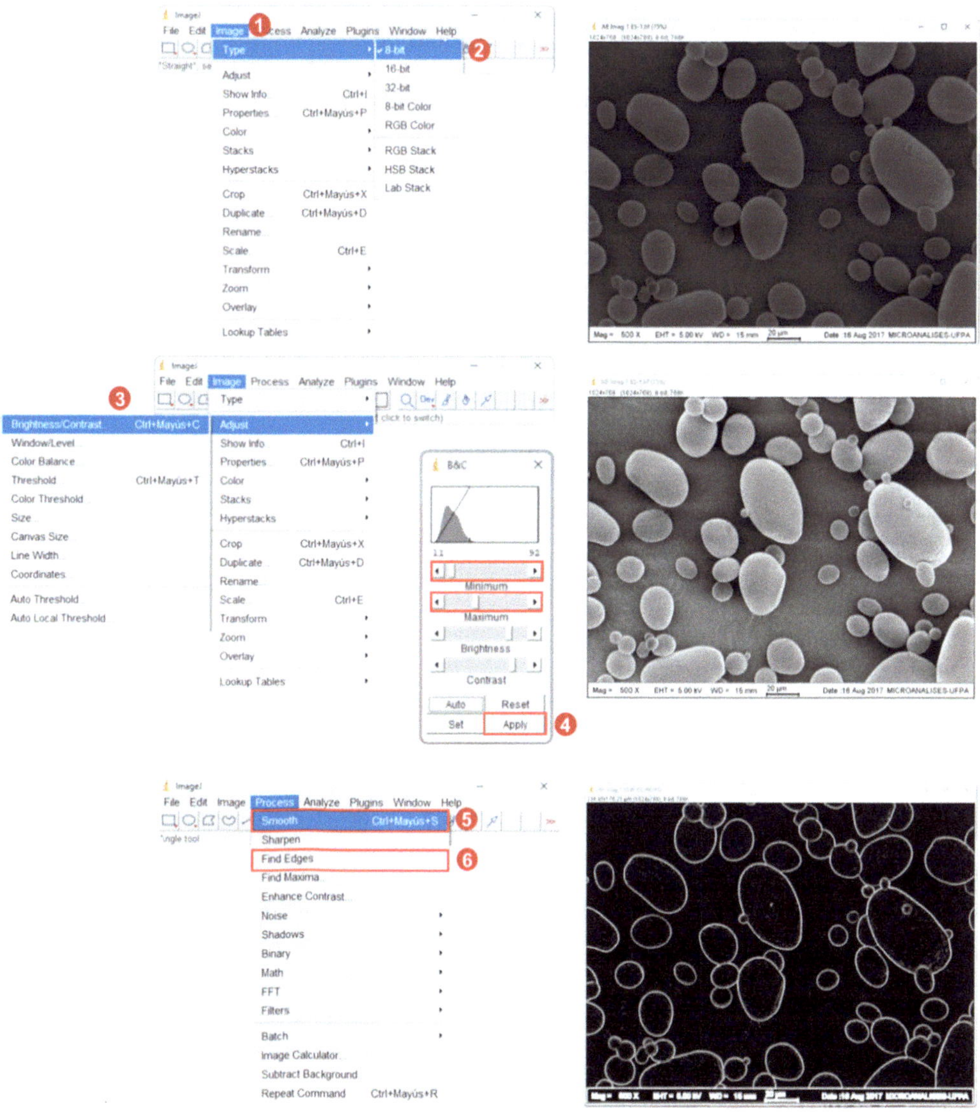

Fig. 4 Prepare the image for accurate particle selection. Source: Authors' collection, 2025

  - Enhance contrast so granules stand out from the background: Image ▶ Adjust ▶ Brightness/Contrast (adjust and Apply).
  - Lightly smooth noise: Process ▶ Smooth (or Median).
  - Visualize edges: Process ▶ Find Edges to display detected granule boundaries.

- *Threshold the granules*
  - Go to Image ▶ Adjust ▶ Threshold....
  - Move the sliders until the binary mask includes the entire interior of each granule and excludes the background.
  - Click Apply to create the binary image as shown in Fig. 5. *Check:* there should be no holes inside granules.

- *Separate touching/overlapping granules*
  - Use Process ▶ Binary ▶ Watershed to split adjacent particles along the narrowest necks as shown in Fig. 6.
  - If a granule is oversegmented, this can be corrected during the selection/analysis stage (e.g., by editing ROIs in the ROI Manager).

3.3 Analysis of the Main Parameters: Area, Perimeter, Length, Circularity

- *Open the ROI Manager and select granules*
  - Go to Analyze ▶ Tools ▶ ROI Manager....
  - With the Wand (tracing) tool, click each granule to select it, and press Add (or key t) to store it in the ROI Manager.
  - If a granule was split during watershed: select all of its fragments as shown in Fig. 7 (Shift-click with the Wand or select

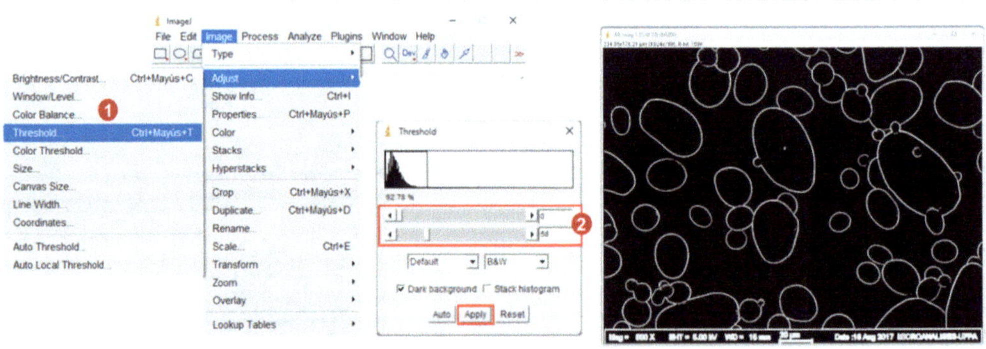

Fig. 5 Threshold the granules. Source: Authors' collection, 2025

their ROIs in the list), then choose More ▶ Combine (OR) to merge them into a single ROI and Update/Add.

- *Measure the selected granules*
  - After adding all granules (one ROI per granule), click Measure in the ROI Manager (Fig. 8).

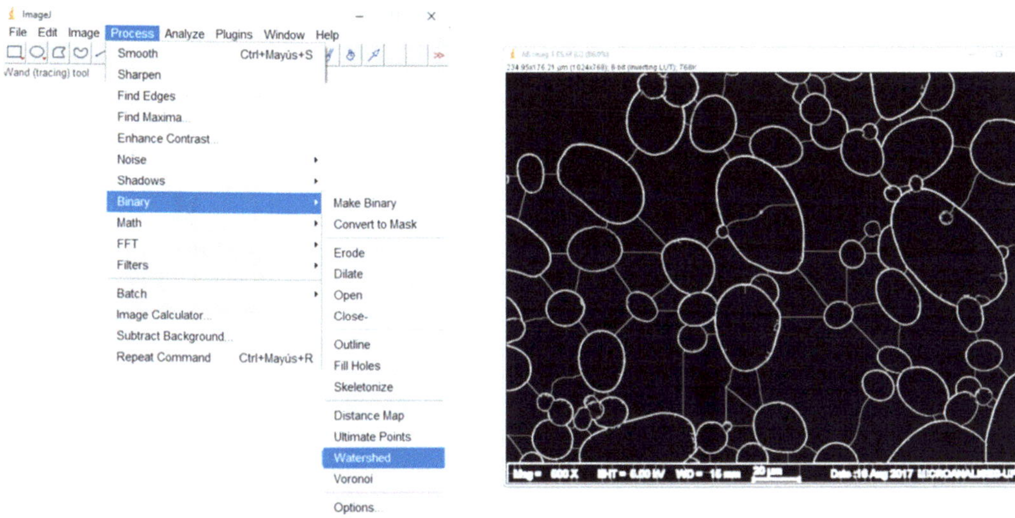

Fig. 6 Separate touching/overlapping granules. Source: Authors' collection, 2025

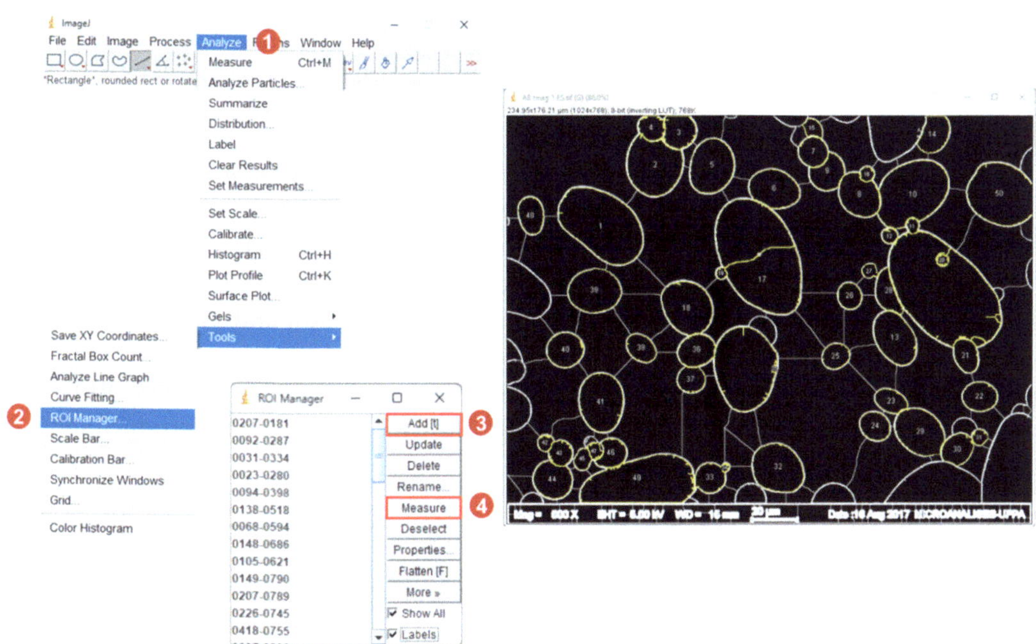

Fig. 7 Analysis of the main parameters of starch granules. Source: Authors' collection, 2025

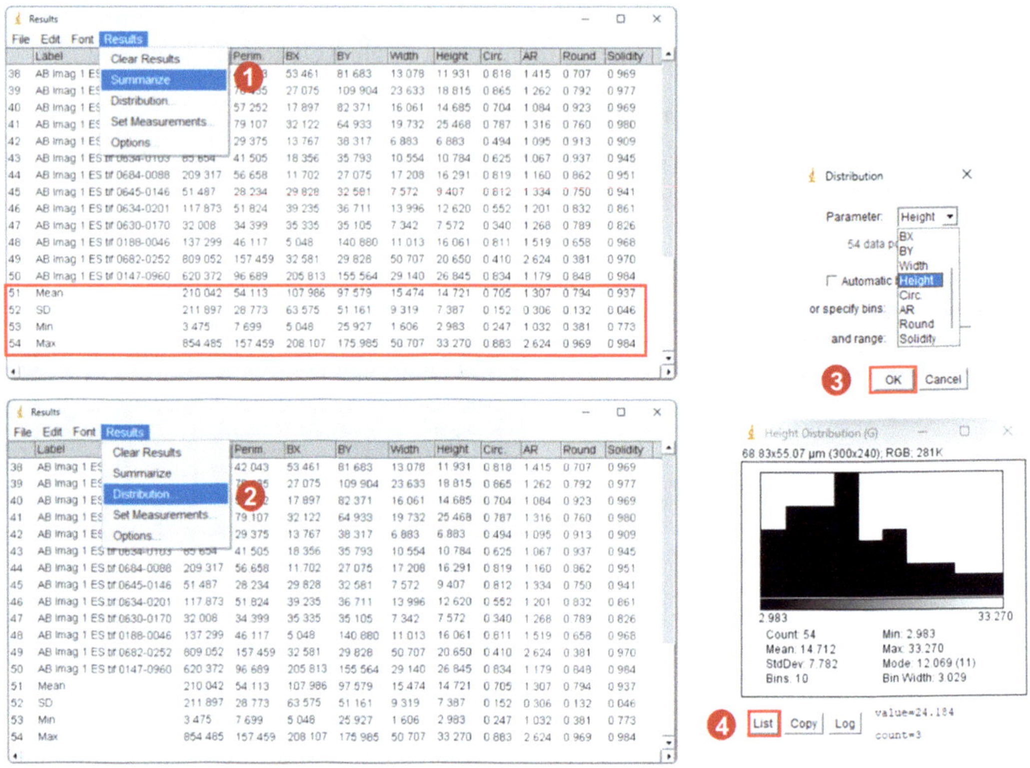

Fig. 8 Measurement the selected granules. Source: Authors' collection, 2025

  - The Results window will list the metrics you enabled in *Set Measurements* (e.g., area [μm^2], perimeter [μm], and shape descriptors such as circularity, roundness, solidity)[1,2,3]

- *Summaries and distributions*
  - To obtain descriptive statistics (count, mean, SD, min, max, etc.), choose Results ▶ Summarize.
  - To plot a size distribution, choose Results ▶ Distribution… , select the column of interest (e.g., Feret, ECD, or Area), define bin number or bin width, and generate the histogram/cumulative curve.
  - Export tables via Results ▶ Save as (.csv); save the ROIs with ROI Manager ▶ More ▶ Save (.zip) for reproducibility.

4 Notes

1. Length can be reported as Feret (max) or as the major axis from *Fit ellipse*; width as MinFeret or minor axis.

2. Circularity = 4π·Area/Perimeter^2 (1.0 = circle). Solidity = Area/Convex area (requires *Shape descriptors*).
3. Inspect overlays and remove ROIs touching the image border or obvious artifacts before final measurements.

References

1. Tester RF, Karkalas J, Qi X (2004) Starch—composition, fine structure and architecture. J Cer Sci 39(2):151–165
2. Pérez S, Bertoft E (2010) The molecular structures of starch components and their contribution to the architecture of starch granules: a comprehensive review. Starch-Stärke 62(8):389–420
3. Singh N, Singh J, Kaur L, Sodhi NS, Gill BS (2003) Morphological, thermal and rheological properties of starches from different botanical sources. Food Chem 81(2):219–231
4. Goldstein JI, Newbury DE, Echlin P, Joy DC, Lyman CE, Lifshin E, Sawyer L, Michael JR (2003) Scanning electron microscopy and X-ray microanalysis, 3rd edn. Springer
5. International Organization for Standardization (2014) ISO 13322-1:2014—Particle size analysis—Image analysis methods—Part 1: Static image analysis methods. ISO
6. Schneider CA, Rasband WS, Eliceiri KW (2012) NIH Image to ImageJ: 25 years of image analysis. Nat Methods 9(7):671–675
7. Schindelin J, Arganda-Carreras I, Frise E, Kaynig V, Longair M, Pietzsch T et al (2012) Fiji: an open-source platform for biological-image analysis. Nat Methods 9(7):676–682
8. Russ JC, Neal FB (2016) The image processing handbook, 7th edn. CRC Press
9. Oliveira LC, Rosell CM, Steel CJ (2015) Effect of the addition of whole-grain wheat flour and of extrusion process parameters on dietary fibre content, starch transformation and mechanical properties of a ready-to-eat breakfast cereal. Int J Food Sci Technol 50(6):1504–1514

Check for updates

Chapter 16

Oscillatory Rheology for Characterization of Starch Gels

Kivia Mislaine Albano, Sungil Ferreira, Poliana Moser and Vânia Regina Nicoletti

Abstract

Oscillatory rheology enables characterization of viscoelastic properties of starch suspensions and gels, which in turn help to reveal structural changes occurring under processing conditions and to establish relationships between structure, sensory and functional properties, product formulation, and equipment design. In this chapter, we describe different oscillatory tests useful to characterize the structure of starch-based material, including strain, frequency, time, and temperature sweeps, as well as the three-interval thixotropy tests.

Keywords Viscoelasticity, Mechanical spectra, Storage modulus, Loss modulus, Gelatinization kinetics, Thixotropy

1 Introduction

Starch gels are composite materials of fluid and solid constituents, and their rheological behavior originates from the combination of the individual constituents' response to applied stress or deformation. The morphology, as well as the spatial distribution and orientation of the suspended particles, strongly affect the rheological properties of starch gels that, in addition, are strongly influenced by swelling and gelatinization of the starch granules. Gelatinized starch suspensions may behave as a viscous dispersion or as a gel depending on starch concentration, temperature, and starch botanical source. The influence of starch concentration and temperature is particularly important, taking into account that these materials can be submitted to different conditions during the processing of starch-based products [1–4]. Knowledge of rheological properties helps elucidate structural changes occurring along processing operations and is an important tool to establish relationships between structure, formulation,

Maria Teresa Pedrosa Silva Clerici and Pedro H. Campelo (eds.), *Basic Protocols in Starch Characterization*, Methods and Protocols in Food Science, https://doi.org/10.1007/978-1-0716-5116-2_16,

processing and design [5–7]. These mechanical properties are also relevant for improving quality control in the food industry and for correlating sensory analysis with more objective and reproducible rheological tests [8, 9].

Starch suspensions and gels behave as partly viscous and partly elastic materials, and the apparent viscosity alone is not enough to characterize their rheological behavior due to the great complexity of molecular rearrangements that follow an applied deformation. In these kinds of systems, viscoelastic effects must be taken into account. Oscillatory shear assays are a fundamental tool that allows assessing the viscoelastic properties of starch-based materials and classifying them into strong or weak gel-forming materials [10]. Measurements in the linear viscoelastic domain, such as small amplitude oscillatory shear flow, can yield very useful information on the suspension structure and interactions between the various components. For a small enough strain, the structure of multiphase polymeric systems is expected not to change, and the linear viscoelastic functions are able to describe the equilibrium properties [5].

The rheological tests under low amplitude oscillatory shear, also called dynamic rheological tests, in which the specimen is deformed by a stress varying with time according to a sinusoidal curve, allow calculating the storage modulus (G') and the mechanical dissipation or loss modulus (G'') over a wide range of frequencies. The mechanical dissipation is associated with the energy loss and heating due to friction between molecules. Since a periodic experiment at frequency ω is quantitatively equivalent to tests at time $t = 1/\omega$, it is possible to obtain enough information in short times about the biopolymer suspension behavior at different time scales by running a frequency sweep and determining the material mechanical spectrum [11, 12]. For oscillatory rheological tests, an increasing deformation scan (dynamic strain sweep) is first made to check the linear viscoelasticity region (LVR), keeping the oscillation frequency constant. All subsequent scans are often performed keeping a constant strain in the range of LVR, in which the material functions do not depend on the magnitude of the stress or the intensity of deformation or strain rate [13, 14]. In addition to (i) amplitude and (ii) frequency sweep assays, other oscillatory tests may be useful to characterize the rheological behavior of starch-based materials, including: (iii) temperature sweep, in which amplitude and frequency are held constant while temperature is ramped, being useful to determine starch gelatinization temperature; (iv) time sweep with amplitude, frequency, and temperature being held constant, and storage and loss modulus monitored over time; and (v) thixotropy evaluation that can be carried out by a three-interval thixotropy test (3ITT), in which frequency is held constant while amplitude is stepped from low to high to low in order to monitor structure before, during, and after destructive shear.

2 Material

Different botanical sources are rich in starch, such as corn, potato, cassava, Peruvian carrot, rice, wheat, among many others. The starch may be extracted by the researcher or purchased in the local market to be evaluated regarding its rheological properties and potential applications (*see* **Note 1**). The total solids content of samples must be determined according to the methods recommended by AACC [15] (*see* **Note 2**).

3 Methods

3.1 Gelatinization Procedure

In this chapter, the starch gels are suggested to be prepared according to Albano et al. [16]. Different concentrations of gels are suggested; however, depending on the origin of the material and property to be evaluated, the operator must determine the optimum gel concentration to be used. The starch gels should be prepared using deionized water, taking into account the previously determined starch water content (*see* **Note 2**). Batches of 100 g of suspension can be prepared for different starch and/or concentrations, and the recommended gelatinization procedure is based on the method described by Paterson et al. [17] (Eq. 1). The temperature of the starch suspensions should be instantaneously raised to a predetermined level to guarantee an isothermal gelatinization, increasing the uniformity of granules throughout the sample. For this to be possible, the calculated amount of starch required for each predetermined concentration should be hydrated with deionized water up to a total of 25 g of suspension. This initial suspension is to be left at room temperature, 25°C (T1), for 10 min. To complete 100 g of the final suspension, 75 g of deionized water previously heated to approximately 100°C (T2) should be added to the initial suspension under manual agitation (using a glass rod). By doing so, the suspension temperature at the beginning of the swelling process should be 81.25°C, with the assumption that the starch dispersions and water have similar specific heat. The gelatinization can be completed by heating the suspension for 30 min at 90°C (T3) in a thermostatic bath (*see* **Note 3**) with slight manual agitation to keep uniform temperature inside the sample cup (250 mL glass beaker) (*see* **Note 4**). Finally, the suspension should be cooled to room temperature (25°C) before the rheological measurements.

$$25\mathrm{T1} + 75\mathrm{T2} = 100\mathrm{T3} \tag{1}$$

where T1 is the sample ambient temperature (about 25°C), T2 is the calculated temperature for the water to be added, T3 is the desired final temperature of the starch suspensions (*see* **Notes 5 and 6**).

Fig. 1 Possible scenarios of sample conditioning in the rheometer. In "**a**" there is too much sample, in "**b**" there is not enough sample, and in "**c**" the correct amount of sample is used

3.2 Oscillatory Rheological Measurements

In this chapter, the authors used as a reference a rotational rheometer AR-2000EX (TA Instruments, Delaware, USA) with a serrated (cross-hatched) parallel plate geometry (∅40 mm) (*see* **Note 7**) with a gap of 800 μm, which was adopted based on preliminary results and literature [16]. Different gap values have been reported for different gel concentrations and starch sources, varying from 500 μm [18] to 1 mm [4, 14, 19, 20] (*see* **Note 8**). Alternative rheometers from different manufacturers (Anton Paar, Thermo Fisher Scientific, Malvern Instruments, Hitachi) can be used, as long as they can perform oscillatory assays and have sensitivity to the aimed conditions.

Approximately 2.0 mL of sample can be introduced into the rheometer (the sample volume may vary depending on the geometry and model/brand) using an automatic pipette, transfer pipette or a spatula (depending on the flow behavior of the sample). All samples should rest/stabilize for 2–5 min before measurements. Performing tests in triplicate is advised. The sample should be properly accommodated in the geometry as indicated in Fig. 1. The rheometer accessory solvent trap can be used during measurements to prevent solvent evaporation (*see* **Note 9**). Most of the rheometers allow temperature control by means of a Peltier system adapted to the measurement geometry. The influence of environmental conditions (such as room temperature and relative humidity) can be avoided using a solvent trap.

3.2.1 Dynamic Strain Sweep (Amplitude Sweep Test)

The proper application of oscillatory assays is conditioned on the fact that the assays are done in the linear viscoelastic region (LVR) of the determined material. LVR is evaluated by a strain or stress sweep performed by a variation of the signal's amplitude, keeping a constant frequency [16]. In the LVR, the ratio between stress and strain at any moment or frequency is independent of the applied stress or strain [21].

The value of stress or strain used to evaluate the material's viscoelastic properties is influenced by the material properties and the rheometer sensitivity (*see* **Note 10**). Some examples of conditions reported to determine LVR of starch gels are: stress varying from 0.1 to 22 Pa [7] and 0.1 to 500 Pa [20], strain from 0.001% to 100% [22], all with frequency of 1 Hz. Figure 2 presents an example of an expected result of a strain sweep performed for 10% (w/w) Peruvian carrot starch gel. The strain sweep was applied from 0.001 to 10, with two different values of frequency (0.1 and 1 Hz) at 25°C. When the storage (G') and loss (G'') modulus have a linear behavior, and G' is greater than G'', the system is in the LVR. For the Peruvian

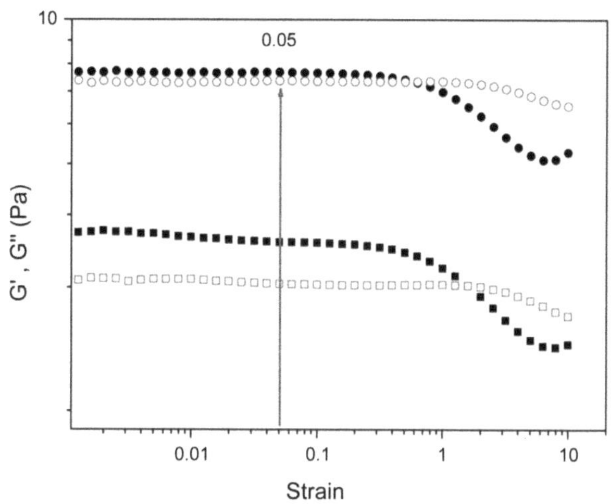

Fig. 2 Linear viscoelasticity region (LVR) of 10% (w/w) Peruvian carrot starch gel at frequencies of 0.1 Hz (*G′* Filled Square, *G′*, Open square) and 1 Hz (*G′* Filled circle, *G′*, open circle), at 25°C [27]

carrot gel tested, the strain considered adequate to be used in further investigations was 0.05 (or 5%). Strong gels remain in LVR up to greater deformations than weak gels [13].

3.2.2 Dynamic Frequency Sweep

The dynamic frequency sweep, which reveals the mechanical spectrum of the material, must be performed in the LVR, as discussed in Sect. 3.2.1. The temperature used may vary, and the frequency range depends on the rheometer's sensibility. Typical reported ranges of frequency are from 0.01 to 10 Hz [7], from 0.1 to 20 Hz [19], or from 0.1 to 25 Hz [23]. Figure 3 shows frequency sweeps in the range of 0.1–10 Hz (0.628–62.8 rad/s) of Peruvian carrot starch gels at different concentrations and temperatures. The assays were performed at 5% strain (in the LVR). A frequency sweep provides information on the material structure and viscoelastic properties, allowing to categorize the gel as diluted solution (G'' larger than G'), concentrated solution (G' and G'' curves intersect in the evaluated frequency range) or gel (G' higher than G''). For concentrated or diluted solutions, G' and G'' vary with the applied frequency, whereas remaining practically constant for gels [13]. Figure 3 shows two types of mechanical spectra, which are associated with different gel structures as discussed by Albano et al. [16].

3.2.3 Dynamic Temperature Sweep

Temperature sweeps are usually used to determine starch gelatinization temperature or to evaluate the gel behavior under process conditions. The temperature may range from 25 to 95°C [19], 40 to 95°C [24], 45 to 100°C [4], 50 to 90°C [16], depending on the expected event to be identified. The heating rate used (°C/min)—typical values are 0.5, 1, 2, or 3°C/min—depends on the information the operator/researcher wishes to access, which may include, for example, gelatinization onset and endset temperatures, and granule breakage due to overheating. In general, starch gelatinization is observed above 40°C and ends around 100°C, considering that above this temperature, there is a decrease in the gel viscosity. After reaching 100°C a decreasing temperature sweep may be performed to evaluate gel syneresis [19, 23]. Temperature sweeps are carried out at constant strain (should be in the LVR) and constant frequency (0.1, 1.0, and 10.0 Hz), with frequency being directly correlated with each material property.

Figure 4 presents a temperature sweep for 6% (w/w) Peruvian carrot starch gel. The gel was prepared at 50°C, as described in Sect. 1, and conditioned on the rheometer at the same temperature (*see* **Note 6**). During the assay, the temperature ranged from 50 to 90°C, with a heating rate of 0.5°C/min and frequency of 0.1 Hz. The

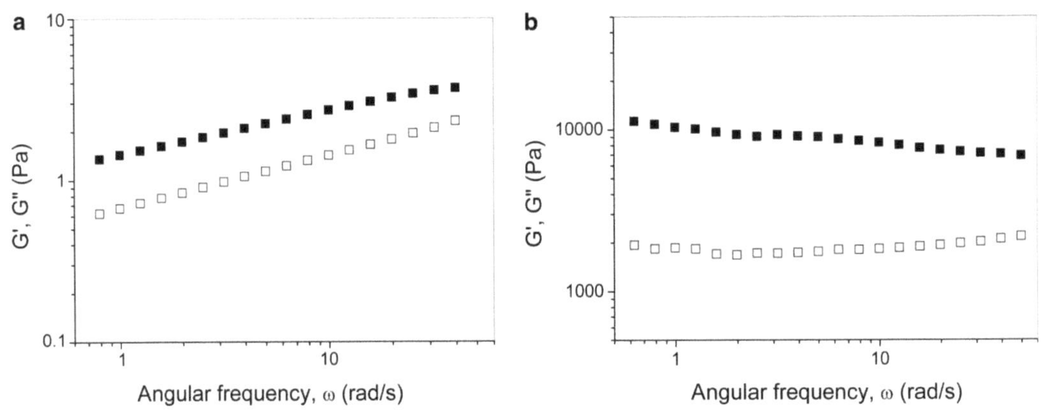

Fig. 3 Storage (*G*′, filled square) and loss (*G*″, open square) modulus of Peruvian carrot starch gel versus frequency: **a** 2% (w/w) at 10°C, and **b** 10% (w/w) at 70°C. Adapted from Albano et al. [16]

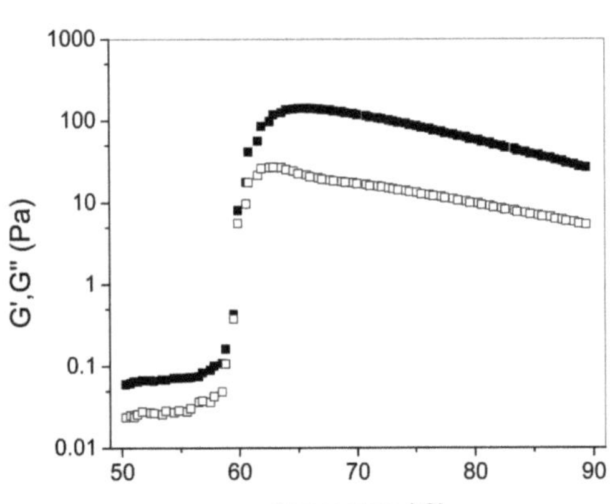

Fig. 4 Evolution of storage (*G*′, filled square) and loss (*G*″, open square) modulus of 6% (w/w) Peruvian carrot starch gel during heating at 0.5°C/min and frequency of 0.1 Hz. Adapted from Albano et al. [16]

gelatinization temperature measured was 61°C, defining gelatinization temperature as the condition corresponding to the crossover of G' and G'' ($G' = G''$) [16, 25].

3.2.4 Dynamic Time Sweep

Time sweeps can be performed to investigate the rheological behavior of starch gels with time, as well as the gelatinization kinetics, around the gelatinization temperature or at a target temperature of interest. When preparing the gel (as per Sect. 1) for time sweep, T3 should be set as the target or desired temperature. Time sweeps should be performed at constant frequency (0.1, 1, or 10 Hz), and the sample should be introduced in the rheometer at the temperature corresponding to the rheological assay (*see* **Note 11**). Figure 5 presents time sweeps of 2% (w/w) Peruvian carrot gel starch at constant temperatures of 59, 62, and 64°C. These temperatures were chosen considering the expected gelatinization temperature (61°C), as presented in Sect. 3.2.3. The time sweeps lasted 30 min (1800 s) at constant frequency (0.1 Hz). The data show that the time necessary for starch gels to attain the plateau of approximately constant G' was longer for lower temperature (59°C) than at higher temperatures (64°C).

3.2.5 Three-Interval Thixotropy Test (3ITT)

The rheological assay known as 3ITT is a tool to simulate different processes applied in the food industry, which involve instant changes in shear stress/shear rate deformation in different magnitudes. The test is composed of a sequence of three time intervals designated as reference, high-shear and regeneration periods. In the first interval,

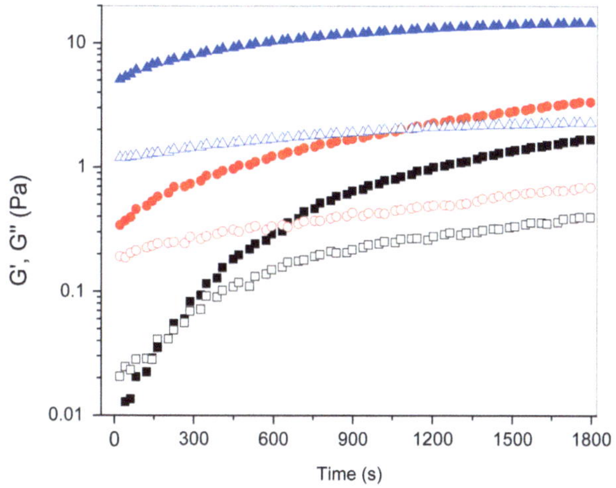

Fig. 5 Profiles of storage (G') and loss (G'') modulus of 2% (w/w) Peruvian carrot starch gel along oscillatory time sweep. The runs were performed at constant temperature of 59°C and frequency of 0.1 Hz (G' filled black square, G'', open black square), 62°C (G' filled red circle, G'', open red circle) and 64°C (G′ filled blue triangle, G'', open blue triangle)

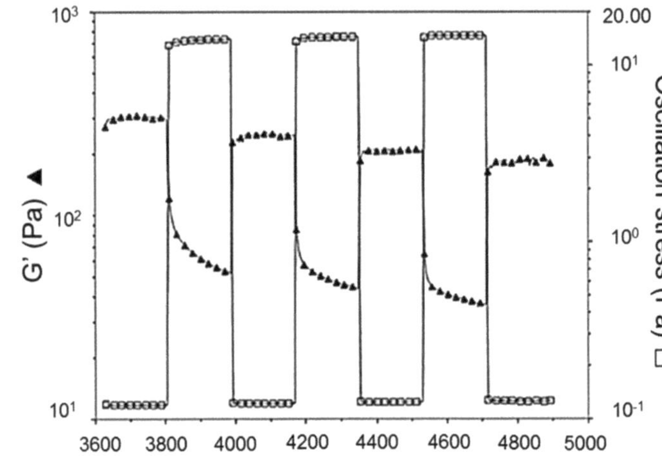

Fig. 6 Structure breakdown and recovery in 10% (w/w) starch paste at 2.5 bar and 100°C. Adapted from Ahuja et al. [7]

low-magnitude shear is applied in the LVR. This is followed by a second interval in which high-shear is performed beyond the LVR. In the third interval, the initial low-shear conditions are again applied in the LVR in order to allow regeneration of the material structure [26].

A few studies reported evaluation of the structure recovery property of starch gels by applying two or more cycles, each cycle with the three mentioned intervals. Chen et al. [4] evaluated starch gels by applying a 1% strain (inside the LVR) for 2 min, followed by 100% strain (beyond LVR) for 2 min, repeating these steps twice (2 cycles), at a constant frequency of 1.6 Hz; their results showed that the physical entanglement networks of the starch samples evaluated were reversible in nature. Ahuja et al. [7] evaluated the structure–recovery behavior of starch pastes at high temperature performing alternate intervals of oscillatory stress: 0.11 Pa for 3 min, followed by 11 Pa for 3 min intending to break the structure, and repeating these steps for three cycles (Fig. 6). It is possible to note that, as the cycles are repeated, the value of G' decreases indicating that the material was not able to fully recover its previous structure. The authors claim that this type of test is particularly important to assess the behavior of a material when it passes through sudden, narrow contraction-expansion during flow.

4 Notes

Gel's or dispersion preparation:

1. During the starch extraction, several factors may affect the characterization properties and yield, such as the region of

harvest, cultivar, soil characteristics, season, weather, extraction method and time, among others. Having this in mind, it is important to perform proximate analysis to help mitigate and explain these differences.

2. After starch extraction and drying, its water content must be measured to ensure the final concentration of the gel to be analyzed. For instance, to prepare 100 g of a 2% starch (w/w) gel from a starch with 12% water content on a dry basis, 2.24 g of starch will be required. Starch must be stored in a hermetically sealed flask under conditions of low humidity and temperature. Otherwise, it is recommended to determine its water content periodically.
3. The operator must be careful not to allow water to evaporate from the sample during gel formulation, since this would change the final concentration. Avoiding that water added to starch reaches the boiling temperature (T2), controlling the temperature of the thermostatic bath (T3), and covering the sample with paraffin film are useful strategies.
4. To produce a completely gelatinized gel, it must be prepared at a temperature varying from 80 to 100°C (which depends on each starch botanical source), being kept in this condition for around 30 min. The dispersion must be stirred slowly and gently (manually or not) and periodically. This avoids breakage of the starch granules and favors gelatinization. Excessive heating may also generate granule' breakage and a change in rheological properties. The complete gelatinization of the granules without their breakage can be observed by optical microscopy.
5. Equation 1 can be used to produce a gel or starch dispersion at a given temperature, depending on the assay to be performed, such as determination of viscous behavior, viscoelastic properties of the gel, completely or partially gelatinized, temperature or time sweep. For that, T3 can always be taken as the final target temperature to evaluate the sample and/or the initial temperature of the assay, i.e., the temperature of the sample to be added to the rheometer. For example, to perform characterization of a starch dispersion at 50°C (T3), T2 will be 58.34°C, being this the water temperature to be added to the dispersion to reach the final target temperature (T3 = 50°C).
6. Once the gel is prepared, it is not recommended to add the sample at high temperature on the rheometer and then decrease the temperature to the assay temperature, except if the assay is to be carried out at high temperature. It is advised to cool the gel at room temperature. If the gel is placed on the rheometer at high temperature, it may leak from the

geometry, or there may be a lack of sample to completely fill the geometry after cooling due to small but significant changes in the sample viscosity and or density, invalidating the run.

Data collection

7. The size of the starch granules is directly related to the starch concentration in the gel and its source. In addition, the size of swollen granules may vary with the gel concentration and the water available in the system. Thus, it is important to determine the proper geometry gap to avoid the breakage of the gelatinized granules in the geometry throughout the rheological assay. It is recommended to use a gap at least ten times larger than the average size of the hydrated starch granules in the sample. Since cone-plate geometry has a small predetermined gap, it is recommended to use parallel plates, in which the gap can be adjusted by the operator, although it is recommended to be as small as possible. The use of cross-hatched parallel plates helps to improve the contact between the geometry and the sample to reduce slip, which is usually caused by large velocity gradients in a thin region adjacent to the wall and causes errors in the rheological measurements.
8. The gap test determination must be performed prior to the oscillatory assays, to ensure no breakage or squeezing of the granules in the geometry, which may occur for high gel concentration. In addition, the gel can leak out of the geometry if the gel concentration is too low. The proper gap may be determined by performing continuous flow assays and using the Power Law (Ostwald-de Waele) model (Eq. 2) to model the shear stress (τ) · shear rate (γ) data. When the flow consistency index (K) does not vary for increasing gaps, it means the starch granules are not being disrupted. The smaller gap after the stabilization of K can be taken as a gap for the oscillatory assays.

$$\tau = K\gamma^n \qquad (2)$$

In Eq. 2, τ (Pa) is the shear stress, γ^n (s^{-1}) is the shear rate, K is the flow consistency index (Pa s^n), n (dimensionless) is the flow behavior index.

9. When performing rheological assays at high temperatures, water may evaporate changing the sample concentration, in addition to forming a dried-polymer film on the sample-air interface that generates friction with the geometry. To avoid this effect, the operator may add a layer of paraffin or oil (sunflower, soy, olive, or mineral oil) around the sample to minimize evaporation. Some equipment provides an accessory, called a solvent trap, that helps to mitigate water evaporation

and the influence of the environment (e.g., temperature and relative humidity). A layer of water is added in a proper place on the geometry, and the solvent trap encloses it. When the temperature increases, the added water evaporates, saturating the air over the sample and minimizing water evaporation. The operator must ensure that the proper amount of water is evenly placed (to avoid overflow during rotation), as well as without uneven water droplets on the geometry (which may interfere with the torque readings), and that there is no physical contact between the solvent trap and the moving geometry.

10. It is important to have as much information as possible regarding the material to be studied and about the rheometer' sensitivity. The literature has several parameters to be used as a reference, such as gelatinization temperature range, concentration, and gap. The rheometer sensibility affects data collection at the lower or higher regions of different assays such as frequencies and/or temperature sweeps, shear rate, stress, or strain range. Some equipment does not have accurate sensitivity in certain conditions, providing inconclusive results.
11. Time sweeps determine the starch gelatinization kinetics at different temperatures, which allows assessing the percentage of starch granules that undergo gelatinization at a certain temperature and time interval. Initially, the starch gelatinization temperature must be determined by carrying out a temperature sweep. The time sweep may be performed at temperatures, for example, 2°C above and below the gelatinization temperature, with each time sweep varying around 0.5°C or more. For example, for a starch with gelatinization temperature of 61.5°C, time sweeps can be performed from 59.5 to 63.5°C, with each sweep varying in 0.5°C, adding up to 9 assays.

References

1. Kim W-W, Yoo B (2009) Rheological behaviour of acorn starch dispersions: effects of concentration and temperature. Int J Food Sci Technol 44:503–509
2. Moreira R, Chenlo F, Torres MD, Glazer J (2012) Rheological properties of gelatinized chestnut starch dispersions: effect of concentration and temperature. J Food Eng 112:94–99
3. Jekle M, Mühlberger K, Becker T (2016) Starch-gluten interactions during gelatinization and its functionality in dough like model systems. Food Hydrocoll 54:196–201
4. Chen H, Xie F, Chen L, Zheng B (2019) Effect of rheological properties of potato, rice and corn starches on their hot-extrusion 3D printing behaviors. J Food Eng 244:150–158
5. Carreau PJ, Lavoie PA, Yziquel F (1999) Rheological properties of concentrated suspensions. In: Siginer DA, De Kee D, Chhabra RP (eds) Advances in the flow and rheology of non-Newtonian fluids. Elsevier, pp 1299–1345
6. Fischer P, Pollard M, Erni P et al (2009) Rheological approaches to food systems. Comptes Rendus Phys 10:740–750
7. Ahuja A, Lee R, Latshaw A, Foster P (2020) Rheology of starch dispersions at high temperatures. J Texture Stud 51:575–584

8. Fischer P, Windhab EJ (2011) Rheology of food materials. Curr Opin Colloid Interface Sci 16:36–40
9. Conti-Silva AC, Ichiba AKT, Silveira AL da et al (2018) Viscosity of liquid and semisolid materials: establishing correlations between instrumental analyses and sensory characteristics. J Texture Stud 49:569–577
10. Marfil P, Telis-Romero J, Nicoletti V (2012) Rheological Behavior of Biopolymer Suspensions. In: Telis VRN (ed) Biopolymer engineering in food processing. CRC Press, Boca Raton, pp 69–110
11. Rao MA (1999) Rheology of fluid and semisolid foods—principles and applications. Aspen Publ., Gaithersburg
12. Mohsenin NN (1986) Physical properties of plant and animal materials, 2nd edn. Gordon and Breach Science Publishers, New York
13. Steffe JF (1996) Rheological Methods, 2nd edn. Freeman Press, East Lansing
14. Precha-Atsawanan S, Uttapap D, Sagis LMC (2018) Linear and nonlinear rheological behavior of native and debranched waxy rice starch gels. Food Hydrocoll 85:1–9
15. AACC (1995) Approved methods of the American Association of Cereal Chemists, 7th edn. St. Paul, Minnesota
16. Albano KM, Franco CML, Telis VRN (2014) Rheological behavior of Peruvian carrot starch gels as affected by temperature and concentration. Food Hydrocoll 40:30–43
17. Paterson JL, Hardacre A, Li P, Rao MA (2001) Rheology and granule size distributions of corn starch dispersions from two genotypes and grown in four regions. Food Hydrocoll 15:453–459
18. Zhu F, Bertoft E, Li G (2016) Morphological, thermal, and rheological properties of starches from maize mutants deficient in starch synthase III. J Agric Food Chem 64:6539–6545
19. Karunaratne R, Zhu F (2016) Physicochemical interactions of maize starch with ferulic acid. Food Chem 199:372–379
20. Villanueva M, De Lamo B, Harasym J, Ronda F (2018) Microwave radiation and protein addition modulate hydration, pasting and gel rheological characteristics of rice and potato starches. Carbohydr Polym 201:374–381
21. Mitchell JR (1980) The rheology of gels. J Texture Stud 11:315–337
22. Molavi H, Razavi SMA (2018) Dynamic rheological and textural properties of acorn (*Quercus brantii* Lindle.) starch: effect of single and dual hydrothermal modifications. Starch/Staerke 70:1–8
23. Wu K, Dai S, Gan R et al (2016) Thermal and rheological properties of mung bean starch blends with potato, sweet potato, rice, and sorghum starches. Food Bioprocess Technol 9:1408–1421
24. Zhu F, Mojel R, Li G (2018) Physicochemical properties of black pepper (*Piper nigrum*) starch. Carbohydr Polym 181:986–993
25. Avanza MV, Puppo MC, Añón MC (2005) Rheological characterization of amaranth protein gels. Food Hydrocoll 19:889–898
26. Toker OS, Karasu S, Yilmaz MT, Karaman S (2015) Three interval thixotropy test (3ITT) in food applications: a novel technique to determine structural regeneration of mayonnaise under different shear conditions. Food Res Int 70:125–133
27. Albano KM (2012) Rheological behavior of Peruvian carrot starch gels (Arracacia xanthorrhiza B.). Master's thesis, São Paulo State University (UNESP), Institute of Biosciences, Letters and Exact Sciences, São José do Rio Preto, SP, Brazil, 113 p

Chapter 17

Mathematical Modeling of RVA Curves and Kinetics of Pasting Starch

Ibrahim Palabiyik, Nasim Kian-Pour, Omer Said Toker

Abstract

The mathematical modeling of the starch pasting curve was used to understand the kinetics of gelatinization, swelling, and retrogradation of starch. Also, it is aimed to attract scientific consideration to the dynamic behavior of the rapid visco analyzer (RVA) curve about process parameters instead of a direct data value on the curve. The RVA curve was divided into four sections, and a modified Hill model for the sigmoidal gelatinization part of the curve, an exponential thinning model, and an Arrhenius thermal equation were used in this study. The parameters of models were used to describe the swelling extent of starch granules, free water-binding capacity, process time, resistance of the starch against shear and heat, swelling rate, and retrogradation rate.

Keywords Kinetics, Hill model, Starch pasting curve, Gelatinization, Swelling rate, Retrogradation rate, Corn starch

1 Introduction

Starch is widely used in pharmaceutical, food, biomedical, and polymer industries as a gelling agent, thickener, stabilizer, and binding agent due to its cheapness, biocompatibility, biodegradability, low price, and nontoxcicity [1, 2]. Starch has discrete semicrystalline granules, with concentric rings of crystalline and amorphous regions. The low solubility of starch granules in water at room temperature, high viscosity, thermal and shear resistance of starch, and its retrogradation often limits its applications in many industries [3]. However, upon exposure to thermomechanical process in the presence of excess water, starch swells extensively up to a certain temperature as a result of penetration of water into the granules due to chemical potential gradient and then the starch granules disintegrate [4, 5]. This irreversible phenomenon is known as gelatinization, or starch pasting. In the gelatinization process, water penetrates the granules and creates a new hydrogen bond between the hydrogen atoms of

Maria Teresa Pedrosa Silva Clerici and Pedro H. Campelo (eds.), *Basic Protocols in Starch Characterization*, Methods and Protocols in Food Science, https://doi.org/10.1007/978-1-0716-5116-2_17,

water molecules with the hydroxyl group of the glucan chain (especially amylopectin chain), which causes starch granules to swell. During the swelling process on the microscopic scale, due to cleavage of weak intermolecular bonds by heat and shear force, the amylose and short-chain amylopectin molecules leach out from the starch granules into the continuous phase and melting of crystallites generally takes place at a temperature of 60–70°C and sometimes rupture in the structure of starch granules takes place. Gelatinization increases the water-binding capacity and, therefore, at a certain temperature, the starch granules dissolve into the water [6, 7]. During this process, the overall viscosity of starch increases, which has a direct relation with the volume fraction of swollen starch granules in the suspension. Therefore, pasting properties have an important effect on the functionality of starch in industrial applications [5, 7].

Between the different techniques for analysis of pasting curves-pasting curves, the rapid visco analyzer (RVA) is one of the most used techniques to study the behavior of starch at the different stages of heating, shearing, and cooling. This method measures the viscosity of starch slurry during the processing time, and the experimental data can be used to determine the pasting parameters of starch. However, only a few authors worked on the mathematical modeling of starch pasting curves [1, 2]. Instead, the traditional methods for comparing data only according to the pasting parameters are inconclusive. For instance, various studies measure the peak viscosity from the RVA curve, while during gelatinization, the rate at which the peak viscosity is attained is rarely studied due to the lack of a mathematical model. However, the modeling is crucial for comprehension of the kinetics of gelatinization, the kinetics of granule swelling, estimating the process times and selecting suitable starch for food products. Therefore, the aim of this work is to apply rational models to analyze each step of a typical starch RVA curve by suitable empirical models. These models will shed light on understanding the relevance of the starch constituents and pasting behavior, and comparing data. Also, it is valuable for process design and for developing food with the desired quality.

2 Material

Corn starch was supplied by local suppliers.

3 Method of Mathematical Modeling of RVA Curve

3.1 Pasting Analyses

1. Prepare starch slurry: Mix starches with water to obtain 14% (w/w) concentration.

2. We used a Rapid Visco Analyzer (RVA) (Perten RVA 4500, Australia) for the determination of pasting properties.
3. Set the initial temperature to 50°C and mix the slurry at 960 rpm for 10 s.
4. Reduce the shearing speed to 160 rpm.
5. Increase the temperature from 50°C to 95°C in 4 min.
6. Continue mixing at 160 rpm and keep the sample at 95°C for 5 min.
7. Reduce the temperature from 95°C to 50°C at a rate of 15°C/min.
8. Record the pasting curve (Fig. 1).

3.2 Mathematical Modeling of the RVA Curve

Divide the pasting curve into four parts (Fig. 2).

3.2.1 First Part

It is placed between the start of the curve (the point at which temperature increased) up to peak viscosity (PV). Fit the experiment data of the "S" shape (sigmoid) curve in this part to a logistic function such as modified Hill model (Eq. 1) [2].

$$V = \frac{V_{\text{peak}} t^S}{R^S + t^S} \tag{1}$$

where V is the viscosity, V_{peak} is peak viscosity (cP), R is the time to reach 50% of peak viscosity (s), t is the time (s), and S is the starch coefficient (*see* **Notes 1–4**).

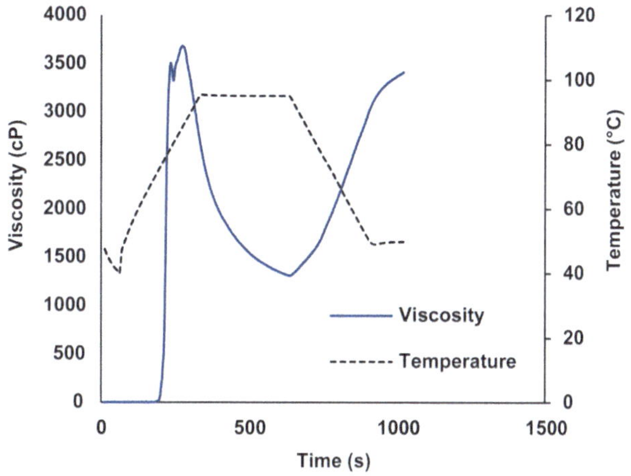

Fig. 1 RVA curve of corn starch

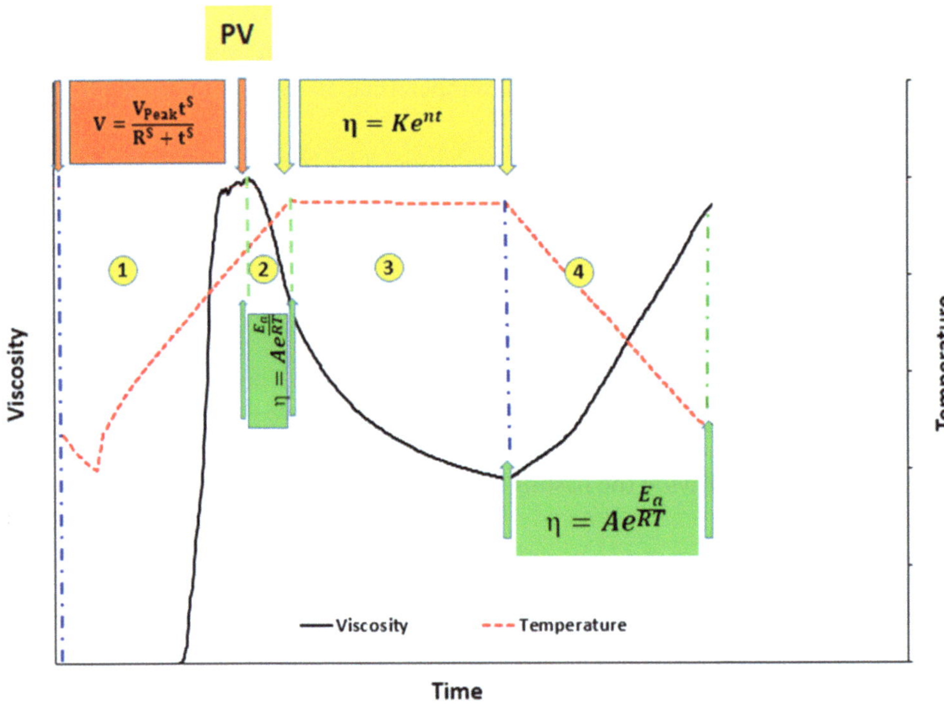

Fig. 2 Schematic diagram of starch RVA curve and the applied models for each part of the curve

3.2.2 Second Part

The second region of the curve starts from the peak viscosity until the start point of temperature stability (95°C). The relation between viscosity and temperature at the constant shear rate in this part was modeled by the Arrhenius equation (Eq. 2).

$$\eta = A_0 \exp\left(\frac{E_a}{\mathrm{RT}}\right) \tag{2}$$

where η is viscosity (cP), E_a is activation energy (J/mol), A_0 is the constant parameter of the Arrhenius model, R is gas constant (8.314 J/mol K) and T is the temperature (K) [1] (*see* **Note 5**).

3.2.3 Third Part

The third part of the RVA curve is the zone in which starch gel is exposed to a constant temperature and shear rate. The exponential model was selected to express the change of starch viscosity with time (Eq. 3)

$$\eta = K \exp(nt) \tag{3}$$

where η is viscosity (cP), K, and n are model parameters, and t represents the processing time at the third region of the curve (s) (*see* **Note 6**).

Table 1
First part of curves: modified Hill model parameters determined for modeling the RVA curve of starch

First region of the RVA curve	Description	Model parameters	Results
$V = \frac{V_{peak} t^S}{R^S + t^S}$	Peak viscosity (cP)	V_{peak}	3743
	The time to reach 50% of peak viscosity (s)	R	243.2
	The starch coefficient	S	4.631
	The coefficient of determination	R^2	0.9754

3.2.4 Fourth Part

The fourth part is the region between the beginning of a decrease in temperature from 95°C until 50°C, and in this part, the relation between viscosity and temperature at the constant shear rate was defined by the Arrhenius equation (Eq. 2) (*see* **Note 7**).

4 Notes

1. The change of viscosity with time in the first part of the curve was fitted to the modified Hill equation by the nonlinear regression method, and the kinetic parameters of V_{peak}, R and S were determined (Table 1). The most important phenomena in this region are the swelling of starch granules and an increase in viscosity.
2. The kinetic parameter of V_{peak} can use for comparing the capacity of starch in binding with free water or it can represent the swelling extent of starch granules. The swelling extent can be affected by temperature and it is higher at the temperatures higher than the gelatinization temperature [5]. The V_{peak} of our experiment was in the agreement with another study [1] but it was slightly higher than the results of [2] which may be related to the origin of the starch.
3. Generally in the literature, the peak viscosity calculated in this region but the how quick PV of the starch slurry was reached, is not known. Therefore, mathematical modeling of this part of the curve is necessary for the determination of swelling rate and resistance of starch against external conditions such as heat and shear force which can represent by the R-value. This parameter shows the necessary time to reach 50% of peak viscosity during swelling phenomena. The R was determined as 250 s for corn starch by the [2] which slightly was higher than our results. They showed that the wheat starch with the R-value of 327 s, was the most resistant starch against heat and shear force and the wheat starch granules show the resistance to swelling which

may be related to their high amount of phospholipid content compare with pea, corn, waxy maize, and Tapioca starch. Also, the author concluded that the existence of stronger hydrogen bonds in wheat starch is another reason for the high R-value.

4. The S value is the starch coefficient and represents the impact of internal factors (such as AM/AP ratio, granule structure, damaged granules) or external factors (such as heating and shear rate) on the swelling rate of starch. The $S = 1$ will show that initial penetration of water into the starch granules does not influence the diffusion of the remaining water molecule in the environment into the granules, however, $S < 1$ means initial swelling by the diffusion of water molecules into the granules caused a decrease in the affinity of starch to absorb more water and it has a negative effect on the swelling. The $S > 1$ represents that, initial diffusion of water molecules into granules positively modifies the entering of remaining water molecules into granules which increases the swelling rate. Besides, for significant swelling, the temperature of the slurry should be higher than the gelatinization temperature [5]. The S value estimated in current research is higher than the 1 value which is in agreement with that found by other studies [1, 2].
5. The shearing force at the second region of the RVA curve was constant as the first part of the curve, while the temperature of gel continuously increased until reached the maximum temperature of 95°C. Therefore, the dominant effect of temperature on the viscosity of starch was modeled by the Arrhenius equation. Same as the first part of the curve, the temperature increased while viscosity starts to decrease from peak viscosity which may be related to the effect of thermal treatment on the rising distance between starch molecules which decreased intermolecular force and subsequently decreased the viscosity [8]. Furthermore, thermal treatment along with shearing may cause damage and rupture in the structure of swollen granules which lead to a drop in viscosity [9]. The model parameters of

Table 2
Second part of curves: Arrhenius equation model parameters determined for modeling the RVA curve of starch

Second region of RVA curve	Description	Model parameters	Results
	Constant parameter of the model (cP)	A_0	164
$\eta = A_0 \exp\left(\frac{E_a}{RT}\right)$	Activation energy (kJ/mol)	E_a	25,679
	The coefficient of determination	R^2	0.8621

the Arrhenius equation are shown in Table 2. The activation energy shows the stability of starch against temperature change. Palabiyik et al. [2] reported *Ea* at the second part of the RVA curve were as 33,391 (kJ/mol) for waxy maize starch, 29,632 (kJ/mol) for tapioca starch, and 15,882 (kJ/mol) for pea starch. According to the linear regression of our experimental data, the *Ea* value was 25,679 (kJ/mol), which indicates that corn starch gel has a high stability against temperature change. Karakelle et al. [1] used different heating rates (2.5, 5, and 10°C/min) and mixing speeds (50, 160, 250 rpm) for modeling the RVA curve of waxy, normal, and high amylose maize starch, and the *Ea* values ranged from 9.87 to 39.3 (kJ/mol). Palabiyik et al. [2] showed a very good correlation between the amylose content of different starches and the *Ea* value in the third region of the curve, and they revealed that increasing in amylose content can lead to decrees in *Ea* value and increase the thermal stability of starch which can be related to the creation of complex structure among leached amylose and starch granules. Various authors studied the effect of different gelatinization temperatures (75°C and 95°C) on the pasting properties of tapioca starch [10], and the effect of traditional RVA temperature 95°C and high-temperature RVA (140°C) on the pasting properties and viscosity of high-amylose wheat starch gel [11]. The mathematical modeling of the RVA curves is a valuable method for understanding starch behavior in terms of kinetics of gelatinization, granule swelling rate, stability of starch gel against heat and shear change, specifying the process times, and selecting the suitable starch for the food product.

6. In the third zone of the RVA curves, the decreasing behavior of starch gel was fitted to an exponential equation via linear regression analyses, and the model parameters were detailed in Table 3. The model constants were determined from the slope of $Ln\eta$ against time. During this period, swollen starch gel was exposed to the maximum temperature of 95°C for 5 min, and due to the effect of simultaneously high temperature and shear,

Table 3

Third part of curves: exponential equation model parameters determined for modeling the RVA curve of starch

Third region of RVA curve	Description	Model parameters	Results
	Parameter of model (cP)	K	2406
$\eta = K \exp(nt)$	Parameter of model (n * 100)	n	0.33
	The coefficient of determination	R^2	0.9661

the granules were ruptured and lost their integrity; therefore, the amount of damaged starch will rise during this period, which leads to a decrease in the viscosity. The *K* value was related to the viscosity, and the *n* value gives information about the internal structure of granules. The *n* value can be an indicator to compare the damaged starch content, and a high level of *n* represents that the granules collapse easily, and it shows the fragile and brittle structure of swollen granules in the gel against the shear force. Therefore, an increase in the amount of damaged starch can result in a high amount of *n* value, which leads to decreases in the stability of starch gels against applied shear force, and it needs to be for the formulation or process condition of food products. Karakelle et al. [1] reported that as the shearing rate increased from 50 to 250 rpm, the *n* value of the normal maize starch increased. It was stated that the highest amount of *n* of waxy maize starch compared with corn, pea, wheat, and tapioca starch was related to its high level of damaged starch [2].

7. In the fourth part of the RVA curve, a constant shear rate was applied to the sample and the starch gel was cooled from a maximum temperature (95°C) to 50°C. As the change of the temperature is the dominant factor in changing viscosity, similar to the second zone, the Arrhenius equation was fitted with experimental data and the results were summarized in Table 4. In the fourth part of the RVA curve, *Ea* represents the retrogradation rate. However, without modeling of RVA curve, setback viscosity (SV) was used in literature to compare the retrogradation rate of different starches. In the gelatinization step, amylose and short-chain amylopectin molecules are leached from the starch granules. Therefore, cooling of starch gel during the fourth step causes an interaction to take place between the leached amylose and the amylopectin chains, which are present in the swollen granules, and it causes recrystallization of starch molecules. Due to recrystallization, a rigid structure

Table 4

Fourth part of curves: Arrhenius equation model parameters determined for modeling the RVA curve of starch

Fourth region of RVA curve	Description	Model parameters	Results
	Constant parameter of the model (cP)	A_0	81
$\eta = A_0 \exp\left(\frac{E_a}{RT}\right)$	Activation energy (kJ/mol)	E_a	21,095
	The coefficient of determination	R^2	0.9938

forms in the gel, and the hardness of cooked food will increase. An increase in the retrogradation rate can be reflected in the high value of *Ea*, which may be related to the amount of free amylose in the gel. In our experiment, the *Ea* in this part of the RVA curve was 21,095 kJ/mol. Palabiyik et al. [2] showed that *Ea* of wheat starch was lowest among the pea, corn, waxy maize, and tapioca starch, while the minimum amylose content and SV values belonged to the waxy maize starch. They concluded that the lowest amount of *Ea* in the wheat starch may be related to the high amount of phospholipid content of wheat starch [2], which forms a binary starch–lipid complex that causes a decrease in the retrogradation rate, *Ea* value and recrystallization rate of starch. Therefore, the SV value may be used to explain the retrogradation extent, while *Ea* may represent the retrogradation rate.

References

1. Karakelle B, Kian-Pour N, Toker OS, Palabiyik I (2020) Effect of process conditions and amylose/amylopectin ratio on the pasting behavior of maize starch: a modeling approach. J Cereal Sci 94:102998
2. Palabiyik İ, Toker OS, Karaman S, Yildiz Ö (2017) A modeling approach in the interpretation of starch pasting properties. J Cereal Sci 74:272–278
3. Liu T, Ma Y, Xue S, Shi J (2012) Modifications of structure and physicochemical properties of maize starch by γ-irradiation treatments. LWT Food Sci Technol 46(1): 156–163
4. Vamadevan V, Bertoft E (2015) Structure-function relationships of starch components. Starch 67(1-2):55–68
5. Desam GP, Li J, Chen G, Campanella O, Narsimhan G (2020) Swelling kinetics of rice and potato starch suspensions. J Food Process Eng 43(4):e13353
6. Palanisamy A, Deslandes F, Ramaioli M, Menut P, Plana-Fattori A, Flick D (2020) Kinetic modelling of individual starch granules swelling. Food Struct 26:100150
7. Vamadevan V, Bertoft E (2020) Observations on the impact of amylopectin and amylose structure on the swelling of starch granules. Food Hydrocoll 103:105663
8. Xiao Y, Shen M, Luo Y, Ren Y, Han X, Xie J (2020) Effect of Mesona chinensis polysaccharide on the pasting, rheological, and structural properties of tapioca starch varying in gelatinization temperatures. Int J Biol Macromol 156:137–143
9. Li C, Dhital S, Gilbert RG, Gidley MJ (2020) High-amylose wheat starch: Structural basis for water absorption and pasting properties. Carbohydr Polym 245:116557
10. Arslan E, Yener ME, Esin A (2005) Rheological characterization of tahin/pekmez (sesame paste/concentrated grape juice) blends. J Food Eng 69(2):167–172
11. Jin Q, Xu X (2020) Microstructure, gelatinization and pasting properties of rice starch under acid and heat treatments. Int J Biol Macromol 149:1098–1108

Chapter 18

Glycemic Index and Load of Carbohydrates In Vivo Method

Marina Vilar Geraldi, Mário Roberto Maróstica Junior and Cinthia Baú Betim Cazarin

Abstract

Scientific evidence has demonstrated a positive correlation between postprandial glycemia and health status. Glycemic index (GI) is a tool used to determine the incremental blood glucose responses in volunteers after consuming carbohydrate-containing food. It is a valid, standardized and reproducible method for classifying carbohydrate-containing foods and guides healthier food choices to prevent or treat cardiometabolic outcomes. This chapter aims to describe the step-by-step procedure of the "in vivo" assay to classify carbohydrate foods' glycemic index.

Keywords Glycemic index, Carbohydrates, Blood glucose response, Classification of foods, Glycemic response, Glycemic load

1 Introduction

The most common energy source in the human diet is starch; this carbohydrate can present different digestion rates depending on the amylose and amylopectin content in its structure [1]. In this way, three types of starch can be categorized according to susceptibility to digestion: readily digested, slowly digested, and resistant starch (included in the dietary fibers group) [2]. The carbohydrate digestion and intestinal absorption rate promote changes in the blood glucose level, stimulating pancreatic insulin release [3]. In addition to the carbohydrate type, other factors such as the amount of carbohydrate, cooking/food processing (degree of starch gelatinization, particle size, food form, cellular structure), interaction with other food components (e.g., fat, protein, dietary fiber, antinutrients, and organic acids) can influence the carbohydrates digestion and the glycemic response of foods [1].

In 1981, the glycemic index (GI) concept was introduced to classify foods containing carbohydrates according to the postprandial glycemia incremental glycemic response [4]. The GI is a measure

Maria Teresa Pedrosa Silva Clerici and Pedro H. Campelo (eds.), *Basic Protocols in Starch Characterization*, Methods and Protocols in Food Science, https://doi.org/10.1007/978-1-0716-5116-2_18,

of the blood glucose-raising ability of the carbohydrate in different foods defined as the incremental area under the glycemic response curve (iAUC) after consumption of a test food portion containing 50 g (it is possible to use 25 g) of available carbohydrate expressed as a percentage of the iAUC response to the same amount of carbohydrate from a reference food taken by the same subject (see equation 1 below). Glycemic index is a qualitative characteristic intrinsic to a specific food and can be used as an important tool to predict the body's glucose response after consuming a single carbohydrate-containing food [5]. In this way, the index classifies the foods as Low GI (≤55), Medium GI (56–69), or High GI (≥70) based on the incremental area under the curve (blood glucose level) after consuming a portion containing 50g of available carbohydrate [2]. Glycemic load (GL) is estimated by the product of the food GI and the amount of available carbohydrates in a food portion, in this way this value is useful to estimate the response of a meal considering the amount of determined carbohydrate source and its glucose raising ability [6].

Long-term consumption of a carbohydrate-rich diet with high GI and GL is associated with an increased risk of cardiometabolic diseases. Epidemiological data show a positive correlation between GI and some outcomes like diabetes, cardiovascular disease, cancer, and body weight management [7]. In this way, there is some evidence that low GI diets can improve the management of diabetes [8], reduce total and LDL cholesterol [9], reduce coronary heart disease events [10], and reduce the risk of cancer development [11].

In 1997, a Joint Food and Agriculture Organization of the United Nations and World Health Organization Expert Consultation (FAO/WHO) published a report describing the methodology for determining the GI of foods [1], which was reviewed and updated in 2005 by Brouns et al. [12]. Additionally, in 2010, the Organization for International Standardization established a standardized scientific method to determine the GI of foods [13]. During these reviews and updates, additional procedures were introduced to improve accuracy and precision [14].

The performance of the FAO/WHO methodology for determining food GI was assessed by 28 international laboratories in an interlaboratory study, which observed a standard deviation of approximately 9, which can be associated with day-to-day individual variation, type of blood collected (venous or capillary), control reference used (white bread or glucose solution), etc. [15, 16]. The issues identified and procedures to improve accuracy were later added to the ISO method to optimize GI measurement. More recently, these procedures introduced in the ISO method have been assessed in an international interlaboratory study involving three laboratories [14]. The results suggested that the procedures are precise enough to distinguish low-GI food (GI < 55) from high-GI food (GI ≥ 70) with 97%–99% probability.

This chapter will describe and summarize the glycemic index test analysis protocol performed to determine the GI in human volunteers.

2 Material

2.1 GI Protocol

The method described in this chapter for the measurement of GI was adapted from that described by Brouns et al. [12], Wolever et al. [15, 16], and according to the recommended procedures by FAO/WHO [1], and by the International Standards Organization (ISO 26642:2010) [13].

2.2 Definitions

Basic definitions are clarified to facilitate understanding of the terms used in the chapter: *Glycemic response (GR)* is the postprandial change in glucose blood concentration after carbohydrate meal ingestion. *Glycemic index (GI)* measures the blood glucose-raising ability of the digestible carbohydrate in different foods. *Glycemic load (GL)* is the total amount of carbohydrate in a portion size of food. *Available carbohydrate* is the glycemic carbohydrate food content digested, absorbed, and metabolized, increasing blood glucose levels. *Test food* is the food whose GI is being determined. *Reference food* is the reference carbohydrate source (glucose solution or white bread), having by definition a GI of 100.

2.3 Ethics Committee Approval

All clinical studies must be conducted according to the guidelines laid down in the Declaration of Helsinki, and all procedures involving human subjects must be reviewed, approved, and monitored by an institutional review board. The study protocol's ethical approval is important to ensure that all ethical procedures are followed, respecting participants' rights and safety. The informed consent document should provide enough information to the participant to understand the risks, potential benefits, and alternatives to the study before joining it. Signing the document and providing consent is not a contract. Participants may withdraw from a study at any time, even if the study is not over. Besides that, all clinical trials should be registered at clinicaltrials.gov, a database of privately and publicly funded clinical studies conducted worldwide.

2.4 Subjects

The physiological or pathophysiological statuses of the subject are factors that concern the study design and need to be considered for precision and comparability of values. Healthy adults without diabetes are recommended for routine testing at a facility. It is not necessary to restrict volunteers according to their ethnicity, sex, age, or BMI. When appropriate methods are applied, the factors such as age, BMI, sex, and insulin resistance are not related to food GI variations in subjects [16, 17].

A minimum of ten subjects is required to determine the GI of foods and provide a reasonable degree of power and precision [1, 13]. To improve the power or to detect small differences, more subjects are needed.

Subjects may be instructed about moderate restriction before each measurement trial [12, 16]. There is no need for rigorous restrictions, such as prohibiting any physical activity before the test and providing subjects with a standardized meal the previous night. Subjects should avoid unusual exercises and can follow their activity routine the day before the test. The individual should be advised to eat a typical meal of their choice, and repeating that meal the night before each test day should be encouraged. Smoking should not be allowed during the test. These recommendations only apply to the measurement of the glycemic index. When the study's objective is other metabolic investigations in addition to the glycemic response, other specific restrictions can be applied.

2.5 Reference Food

For GI measurement, the carbohydrate portion of the reference and testing food should be based on available carbohydrates. The reference food could be a glucose solution containing 50 g of glucose (50 g of anhydrous glucose or 55 g dextrose dissolved in 250 mL water) or 50 g available carbohydrate from white bread. When choosing white bread as a reference food, it is necessary to standardize the preparation conditions and calibrate against glucose. As the area under the curve (AUC) of the reference food is used as the denominator of the test food, it is recommended to repeat the trials of the reference food to obtain at least two values in each subject to reduce the variation of the mean.

2.6 Testing Food

The portion of test food should contain 50 g of available carbohydrate. If the test food has low or moderate carbohydrate content, it is justified to reduce the portion size to 25 g of available carbohydrate to avoid an unrealistically large meal size. In this case, the reference food must also be adjusted to 25 g [12].

The available carbohydrate should be analyzed with approved specific analytical methods whenever possible to set the portion size. Appropriate methods are available to analyze the carbohydrate in foods, i.e., fibers, total starch, mono and disaccharides for available carbohydrate calculation [18]. The available carbohydrate content can be calculated according to the equation provided by Brouns et al. [12]: total starch × 1.1—resistant starch × 1.1 + total disaccharides × 1.05 + total monosaccharides—nondigestible sugars = g available carbohydrates/100 g food as eaten. After defining the nutritional content, the test food should be served to each participant in portions containing 50 g (or 25 g) of available carbohydrates.

3 Method

3.1 Experimental Procedures

GI trials feature a crossover design, which means that all subjects undergo all treatments. It is recommended that a randomization list be established before the start of the study. Foods for testing should be randomized in blocks of up to six foods, taking care to use a reference food before and after each block. Also, if there are more than two food samples in the study, the total test duration of GI determination should not exceed four months to avoid seasonal variations [12].

Experimental procedures should be taken at breakfast time after volunteers have fasted overnight (10–12 h). Collection of two fasting blood samples at −10 and −5 min before the test is recommended; these two-time measurements need to be taken to average the glucose concentration to be the baseline (fasting) [13]. Then the reference or testing food should be served together with 250 mL of drink, which can be water, tea, or coffee [16]. Prioritize water use, but the important thing is that the same drink should be used in the test with the sample and the reference. On at least two occasions, they must consume the reference food, and on the other occasions, the different test foods. Participants need to be instructed to consume all the food and water within 12 min. Further blood samples are collected at 15, 30, 45, 60, 90, and 120 min after eating the test meal to estimate the glucose incremental area under the curve compared to the reference. Figure 1 exemplifies the experimental procedures for glycemic index measurement.

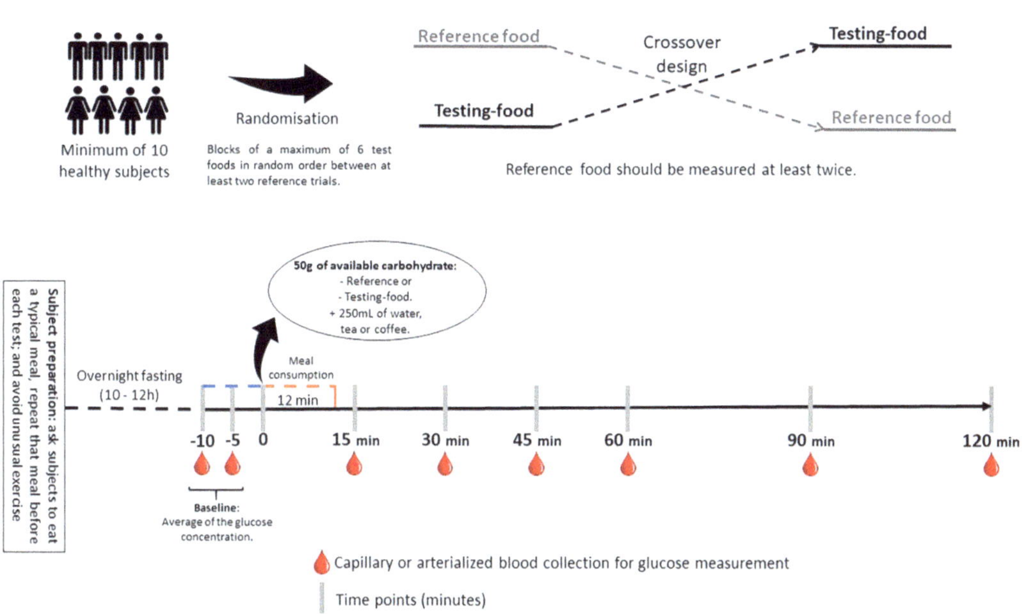

Fig. 1 Experimental procedures for glycemic index measurement. Source: own authorship created using PowerPoint (PowerPoint, Microsoft Office 365®, Redmond, WA, USA) software

Blood samples can be collected by finger-pricking or arterial–venous blood. The sensitivity of glucose measurement is greater using capillary blood sampling; however, the collection from a forearm vein is also recommended. If you decide to use the arterial–venous blood, it is advised to use a winged butterfly system to collect the blood to ensure a painless process for the participants. Glucose concentration can be determined in whole blood, plasma, or serum as long as the analytical method used has a CV < 3.6% [13].

3.2 Calculation of Glycemic Response Measurements

The GI should be calculated using the incremental area under the curve (iAUC) between 0 and 120 min, ignoring the area below the baseline, applying the trapezoidal rule [1]. The average of the two fasting values (−10 and −5 min) should be used as the fasting concentration. The GI of a test food in each participant is expressed as a percentage of the mean iAUC for the reference food consumed by the same subject with an equal available carbohydrate portion, and should be calculated as follows [1]:

$$\mathrm{GI} = \frac{(\mathrm{iAUC\ for\ the\ test\ food})}{(\mathrm{iAUC\ of\ the\ reference\ food})} \times 100$$

The glycemic load (GL) is the product of GI and the total available carbohydrate content calculated according to the following formula [1]:

$$\mathrm{GL} = \frac{(\mathrm{GI} \times \mathrm{available\ carbohydrate})}{100}$$

The GL can have different modes of expression, for example, g per serving, g per 100 g food, and g per 1000 kJ or 1000 kcal. Outliers (defined as values outside the range of mean ± 2SD) are recommended to be excluded [13, 16]. The GI of each test food should be calculated as the mean of all volunteers. According to the ISO method, the mean within-individual coefficient of variation of iAUC cleared by repeated reference oral glucose tests (termed reference CV) must be ≤30% for a valid GI measurement [13]. The reference CV is calculated using the mean, SD, and CV (100 × SD/mean) of the glucose iAUC values obtained from the repeated tests of a reference food (at least two) for each volunteer, and the mean of the resulting values is the reference CV. The GI can predict the glycemic response after the consumption of different GI foods (Fig. 2). Depending on the GI values, foods can be classified as Low- (≤ 55), Medium- (56–69), or High-GI (≥70).

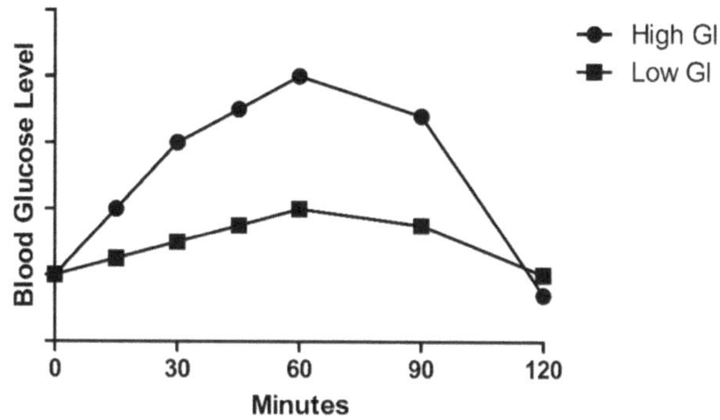

Fig. 2 Glycemic response after consumption of a low- and high-GI food. Source: own authorship created using GraphPad Prism 5.0 (GraphPad Software, Inc., La Jolla, CA, USA) software

4 Notes

4.1 Summary of Recommendations for a GI Trial

4.1.1 Subject's Recommendations

- Number: minimum of 10 healthy subjects.
- Characteristics: healthy human volunteers are recommended, and no need to restrict the age, sex, BMI, or ethnicity.
- Subject preparation: The evening before a test, ask subjects to eat a typical meal, repeat that meal before each test, and avoid unusual exercise.

4.1.2 Reference and Test Food

- Reference food: 50 g of available carbohydrate from glucose solution or white bread, and the glycemic curve should be measured at least twice.
- Test-food: 50 g of available carbohydrate.
- Randomization: Blocks of a maximum of six test foods in random order between at least two reference trials.

4.2 Experimental Procedures

- Meal consumption: the reference or test-food can be served with 250 mL of water, tea or coffee, and should be consumed within 12 min.
- Time points for blood collection: −10, −5 (fasting), 15, 30, 45, 60, 90, and 120 min.
- Blood sampling: capillary or arterialized venous blood.
- Glucose measurement: analytical method with CV < 3.6%.

- GI determination: glucose measurement is enough.
- Total duration: not to exceed 4 months.

4.3 Data Analysis

- AUC calculation: incremental AUC, ignoring area under the baseline.
- Outliers: exclude values outside the range of mean ± 2SD.
- GI calculation: mean of the individual ratios.
- Reference CV: ≤30% for a valid GI measurement.

Acknowledgments

This study was financed in part by the Coordenação de Aperfeiçoamento de Pessoal de Nível Superior—Brazil (CAPES, Finance Code 001), Conselho Nacional de Desenvolvimento Científico e Tecnológico (CNPq, grant numbers 421857/2018-8, 403328/2016-0; 301496/2019-6), and FAPESP (2015/50333-1; 2018/11069-5; 2015/13320-9, 2019/13465-8)." MRMJ acknowledges Red Iberomericana de Alimentos Autoctonos Subutilizados (ALSUB-CYTED, 118RT0543). Marina Vilar Geraldi thanks the FAPESP (grant number 2019/12244-8) for her doctoral fellowship.

References

1. Food and Agricultural Organization (1998) Carbohydrates in human nutrition. Report of a Joint FAO/WHO Expert Consultation. World Health Organization
2. Aguiar LM, Cazarin CBB (2021) In vitro and in vivo methods to predict carbohydrate bioaccessibility. Curr Opin Food Sci 42:69–75
3. Hamaker BR (2021) Current and future challenges in starch research. Curr Opin Food Sci 40:46–50
4. Jenkins DJ, Wolever TM, Taylor RH et al (1981) Glycemic index of foods: a physiological basis for carbohydrate exchange. Am J Clin Nutr 34:362–366
5. Wolever TM, Augustin LS, Brand-Miller JC et al (2017) Glycemic index is as reliable as macronutrients on food labels. Am J Clin Nutr 105:768–769
6. Lal MK, Singh B, Sharma S et al (2021) Glycemic index of starchy crops and factors affecting its digestibility: a review. Trends Food Sci Technol 111:741–755
7. Augustin LSA, Kendall CWC, Jenkins DJA et al (2015) Glycemic index, glycemic load and glycemic response: an International Scientific Consensus Summit from the International Carbohydrate Quality Consortium (ICQC). Nutr Metab Cardiovasc Dis 25:795–815
8. Brand-Miller J, Hayne S, Petocz P, Colagiuri S (2003) Low-glycemic index diets in the management of diabetes: a meta-analysis of randomized controlled trials. Diabetes Care 26:2261–2267
9. Goff LM, Cowland DE, Hooper L, Frost GS (2013) Low glycaemic index diets and blood lipids: a systematic review and meta-analysis of randomised controlled trials. Nutr Metab Cardiovasc Dis 23:1–10
10. Mirrahimi A, de Souza RJ, Chiavaroli L et al (2012) Associations of glycemic index and load with coronary heart disease events: a systematic review and meta-analysis of prospective cohorts. J Am Heart Assoc 1(5):e000752
11. Turati F, Galeone C, Gandini S et al (2015) High glycemic index and glycemic load are associated with moderately increased cancer risk. Mol Nutr Food Res 59:1384–1394
12. Brouns F, Bjorck I, Frayn KN et al (2005) Glycaemic index methodology. Nutr Res Rev 18:145–171
13. Standardization IO for (2010) ISO 26642:2010—Food products—Determination

of the glycaemic index (GI) and recommendation for food classification. Geneva, Switzerland
14. Wolever, Meynier, Jenkins et al (2019) Glycemic index and insulinemic index of foods: an interlaboratory study using the ISO 2010 method. Nutrients 11:2218
15. Wolever TMS, Vorster HH, Björck I et al (2003) Determination of the glycaemic index of foods: interlaboratory study. Eur J Clin Nutr 57:475–482
16. Wolever TMS, Brand-Miller JC, Abernethy J et al (2008) Measuring the glycemic index of foods: interlaboratory study. Am J Clin Nutr 87:247S–257S
17. Lan-Pidhainy X, Wolever TMS (2011) Are the glycemic and insulinemic index values of carbohydrate foods similar in healthy control, hyperinsulinemic and type 2 diabetic patients?. Eur J Clin Nutr 65:727–734
18. McCleary BV, McNally M, Rossiter P et al (2002) Measurement of resistant starch by enzymatic digestion in starch and selected plant materials: collaborative study. J AOAC Int 85:1103–1111

Chapter 19

Prebiotic Activity of Resistant Starches

Beatriz Manfrinato Ferreira, Eliene Penha Rodrigues Pereira and Anderson S. Sant'Ana

Abstract

Resistant starch (RS) is defined as that fraction of carbohydrate that is not digested by digestive enzymes. However, RS can serve as a substrate in fermentation processes of bacteria present in the gastrointestinal tract, especially probiotics. The prebiotic activity of RS can be obtained by microbiological analysis in vitro by probiotic bacteria counts in culture medium enriched with RS or by in vivo analysis by the evaluation of the health parameters of animals or humans after a period of diet rich in RS.

Keywords Probiotic bacteria, Microbiological count, In vitro assay, In vivo assay, Short-chain fat acid, Health parameters

1 Introduction

Carbohydrates can play an important role as a modulating factor on the growth of gut microbiota [1], acting as a prebiotic. Authors have related a tendency of prebiotics use instead of probiotics due to the greater resistance to cross the digestive tract barrier, low costs and safety [1–3].

Resistant starch (RS) is a kind of starch that is not digested in the stomach and small intestine. It can reach the large intestine, where it is used as substrate for microbial fermentation [4]. The fermentation process can benefit the growth of probiotic bacteria, including *Lactobacilli* and *Bifidobacteriaceae* families [5].

Furthermore, bacterial fermentation in the human colon is capable of promoting the production of compounds of great importance for the maintenance of health and the physiological system, called short-chain fatty acids (SCFAs) such as butyrate, propionate, and acetate [4, 6]. Among short-chain fatty acids, butyrate has been most extensively studied and is described as one of the main products resulting from the selective fermentation of resistant starch in the diet by the colonic microbiota, including species like *Eubacteriumrectale, Ruminococcusbromii*, and *Bifidobacterium* [7, 8].

Maria Teresa Pedrosa Silva Clerici and Pedro H. Campelo (eds.), *Basic Protocols in Starch Characterization*, Methods and Protocols in Food Science, https://doi.org/10.1007/978-1-0716-5116-2_19,

Diets containing high levels of fat and sugar can promote harmful changes in the gut microbiome, influencing the intestinal barrier and resulting in a persistent elevation of bacterial lipopolysaccharide (LPS) in the blood plasma. This gut imbalance can contribute to chronic low-grade inflammation in the gut and systemic sites, occasioning the development of diseases, like type 2 diabetes and obesity. The consumption of prebiotics has been reported as a promoter of metabolic benefits, helping to treat the symptoms and progression of these diseases in order to improve insulin resistance and glucose homeostasis, with these characteristics being outstandingly associated with the consumption of resistant starch [9–11].

Therefore, the prebiotic characterization, as well as its efficiency, can be measured by in vitro microbiological analyzes, followed by physical–chemical analyzes and/or in vivo analyzes with tests on animals or humans.

2 Materials

2.1 Resistant Starch Determination

Quantification of resistant starch in a sample can be done by simulating a digestive process. The undigested fraction is considered the concentration of resistant starch. For this, buffer solutions are used, with pH close to that found in the digestive system, and enzymes also found in the gastrointestinal tract, as described below:

1. 2 mol L^{-1}HCl and 0.5 mol L^{-1}NaOH solutions for adjusting the pH of the buffer solutions.
2. Buffer solution KCl–HCl, pH 1.5.
3. Buffer solution 0.1 mol L^{-1}Tris-maleate, pH 6.9, containing 0.004 mol L^{-1} of $CaCl_2$.
4. Pepsin solution prepared using the following ratio: 1 g of pepsin in 10 mL of KCl–HCl, pH 1.5 buffer solution.
5. Pancreatic α-amylase solution prepared using 40 mg of α-amylase per mL of Tris-maleate, pH 6.9 buffer solution.
6. Solution of 0.1 mol L^{-1} KOH.
7. Buffer solution of 0.4 mol/L sodium acetate, pH 4.75, containing 0.02 mol L^{-1} of $CaCl_2$.
8. Amyloglucosidase solution.
9. Glucose water standard solutions for the construction of a calibration curve.
10. Glucose oxidase–peroxidase determination kit (GOD-PAP).
11. Equipments: bath with agitation at temperatures of 37, 40, and 60°C; centrifuge (3000 × *g*) and spectrophotometer calibrated at 500 nm.

2.2 Prebiotic Effect of Resistant Starch by Probiotic Bacteria Counts

2.2.1 Analysis via Microbiological Count

1. Culture media suitable for the probiotic strains used in the study.
2. One or more probiotic strains, such as *Lactobacillus acidophilus*, *Lactobacillus casei*, *Bifidobacterium animalis*, and *Bifidobacterium lactis*.
3. Microplates.
4. An optical density reader to measure the absorbance at 655 nm.

2.2.2 Analysis via the Growth of Viable Cell Number and Metabolic Activity

1. MRS broth (*see* **Note 4.1**).
2. Different sugar sources: glucose, lactose, fructooligosaccharide, and the resistant starch that needs to be verified for its prebiotic action.
3. One or more probiotic strains, like *Lactobacillus acidophilus*, *Lactobacillus casei*, *Bifidobacterium animalis*, and *Bifidobacterium lactis*.
4. One or more strains of enteric bacteria, like *Escherichia coli*.
5. Saline solution for serial dilutions.
6. MRS Agar. To *Bifidobacterium animalis* enumeration, the MRS agar must be supplemented with 0.5 g $L^{-1}$of l-cysteine·HCl and incubated under anaerobic conditions.
7. M9 broth for the enteric culture activation, supplemented with the same sugar sources mentioned above.
8. Eosinmethylene blue Agar for the enteric culture count.

2.2.3 Short-Chain Fat Acids Quantification

1. Standard acids chromatographic grade: acetic, propionic and butyric.
2. 2-ethylbutyric acid solution containing HCl 5 mol L^{-1}.
3. Filters 0.22 μm.
4. Centrifuge.
5. Gas chromatography with a flame ionization detector (FID).
6. Capillary column (30 m × 0.25 mm × 0.25 μm).

2.3 Prebiotic Effect of Resistant Starch by In Vivo Study—Animals

2.3.1 Diet

1. Baseline diet employed.
2. Weighing scale for animals.

3. Syringes and hypodermic needles for intramuscular injection.
4. Xylazine and ketamine as agents for rodent injectable anesthesia.

2.3.2 DNA Extraction and Microbial Profiling

1. QIAmp DNA Stool kit.
2. IlluminaMiSeq integrated next-generation sequencing instrument.

2.3.3 Short-Chain Fatty Acid (SCFA) Analysis

Perform as described in the item (2.2.3).

2.3.4 Evaluation of Animal Health Parameters

1. Glucometer.
2. Syringes and hypodermic needles for intramuscular injections.
3. Glucose and insulinfor biochemical tests.
4. Blood collection tubes.
5. Centrifuge (3000 × *g*).
6. Commercial kits for serum biochemical analysis of uric acid, creatinine, AST (aspartate aminotransferase), ALT (alanine aminotransferase), blood glucose, total proteins, albumin, triacylglycerol (TG), total cholesterol (TCH), high-density lipoprotein cholesterol (HDL-c), and low-density lipoprotein cholesterol (LDL-c).

2.4 Prebiotic Effect of Resistant Starch by In Vivo Study—Humans

2.4.1 Fecal Sample Collection and 16S rRNA Gene Sequencing

1. Stool collection tubes for isolation of DNA using the OMNIgene-Gut (DNA Genotek) collection kit.
2. Freezer (−20°C).
3. Laboratory scale.
4. MoBio PowerMag Soil DNA Kit for DNA extraction.
5. Illumina MiSeq integrated next-generation sequencing instrument.

2.4.2 Short-Chain Fatty Acid (SCFA) Analysis

Perform as described in the item (2.2.3)

2.4.3 Biochemical Analyzes

1. Gloves, tourniquet, antiseptics, cotton and/or gauze pads, needles, vacutainer tube holders and vacutainer tubes, syringe systems, evacuated blood collection tubes, and adhesive bandages for blood analysis.
2. Centrifuge (1048 × *g*).
3. ELISA (Human Insulin ELISA kit, TNFα, and interleukin 6 (IL6) Human ELISA kits).

4. Commercial kits for blood serum sample analysis of total cholesterol (TC), triglycerides (TG), high-density lipoprotein cholesterol (HDL-c) and glucose.

3 Methods

To study the prebiotic capacity of resistant starches present in food samples, they must initially be extracted from the matrices in order to minimize the possible interferences in the analysis. For the extraction methods, *see* **Notes 4.2 and 4.3**.

3.1 Resistant Starch Determination

The determination of resistant starch in a sample can be done through simulated in vitro digestion, followed by quantification by spectrophotometric method, according to Goñi et al. and Pereira et al. [12, 13].

1. In a 50 mL centrifuge tube, add a 100 mg portion of the dry sample or an equivalent portion for humid samples. Add 10 mL of KCl–HCl buffer solution and mix.
2. Add 0.2 ml of pepsin solution. Mix well and keep in a warm bath at 40°C under constant agitation for 60 min.
3. After the period, allow the sample to cool to room temperature and add 9 mL of Tris-maleate buffer solution.
4. Add 1 mL of α-amylase solution. Mix well and incubate for 16 h in a warm bath at 37°C with constant agitation.
5. Centrifuge the samples for 15 min at 3000*g* and discard the supernatant. Wash the precipitate using 10 mL of distilled water, centrifuge again and discard the supernatant.
6. To the residue, add 3 mL of distilled water and mix well. Add 3 mL of KOH solution, mix and keep under agitation at room temperature for 30 min.
7. Add 5.5 mL of HCl solution and 3 mL of sodium acetate buffer solution.
8. Add 80 μL of amyloglucosidase. Mix well and leave in a warm bath at 60°C under constant agitation for 45 min.
9. Centrifuge the mixture, collect the supernatant and save it in a volumetric flask. Wash the residue 10 mL of distilled water and centrifuge again. Add the supernatant obtained previously. Make up to 25–1000 mL, depending on RS content. Filtration of the samples could be performed instead of centrifugation.
10. Proceed by determining the glucose concentration in the solution obtained. An enzymatic test is recommended. Add 0.5 mL of water (blank), sample and standard into test tubes.

Add 1 ml of the glucose determination kit (GOD-PAP) reagent. Mix well and leave to rest for 30 min in a warm bath at 37°C.

11. Read the absorbance of the samples and standards against a reagent blank in a spectrophotometer at 500 nm. The read should be done between 5 and 45 min after incubation.
12. Calculations: Use the standard curve to calculate the glucose concentration of the samples.
13. The resistant starch concentration of the sample is calculated as mg of glucose × 0.9.

3.2 Prebiotic Effect of Resistant Starch by Probiotic Bacteria Counts

The prebiotic effect of resistant starches can be assessed by observing the growth of a probiotic bacterium, as described by Pereira et al. [13].

3.2.1 Analysis via Microbiological Count

1. Reactive the cryopreserved strain in a culture broth under suitable conditions for the probiotic strain selected.
2. The strain must then be inoculated into a microplate containing a solution of a known concentration of the resistant starch under study.
3. Read the initial optical density by measuring the absorbance at 655 nm.
4. The microplate should be incubated at 37°C for 48 h, with monitoring every 30 min by the optical density.
5. A specific growth curve for each strain should be constructed and its slope calculated in order to compare the results obtained.

3.2.2 Analysis via the Growth of Viable Cell Number and Metabolic Activity

This method can be used to confirm the influence of resistant starch on the growth of probiotic bacteria, according to Pereira et al. [13] and de Albuquerque et al. [14], using MRS broth with different sugar sources. An enteric culture mixture, such as *Escherichia coli* strains, can be used for comparison. According to Zhang et al. [15], if a substance can be metabolized by probiotics but not by an enteric mixture, a positive prebiotic activity score will be obtained.

1. In a tube containing 10 mL of each MRS broth (different sugars and the resistant starch), place 0.2 mL of probiotic bacterium inoculum.
2. Homogenize the mixtures and count the colony-forming units (CFU) using serial dilutions with sterile saline solution and plate on MRS agar. It will be the count at time zero (immediately after inoculation).

3. Incubate at 37°C and repeat the count after 48 h. The viable cells must be expressed as log CFU mL^{-1}. The viable cells can also be monitored by optical density (655 nm).
4. Simultaneously, activate the enteric culture mixture in sterile flasks containing 10 ml of M9 broth.
5. Count the viable colonies at time zero using serial dilutions with sterile saline solution and plate on eosin methylene blue Agar.
6. Incubate at 37°C for 48 h and repeat the count.
7. The prebotic activity can then be calculated according to Eq. 1:

$$\text{Prebiotic activity score} = \left[\frac{\left(\textit{probiotic log log}\frac{CFU}{mL}\textit{on the prebiotic at 48 h} - \textit{probiotic log log}\frac{CFU}{mL}\textit{on the prebiotic at 0 h}\right)}{\left(\textit{probiotic log log}\frac{CFU}{mL}\textit{on glucose at 48 h} - \textit{probiotic log log}\frac{CFU}{mL}\textit{on glucose at 0 h}\right)}\right] - \left[\frac{\left(\textit{enteric log log}\frac{CFU}{mL}\textit{on the enteric at 48 h} - \textit{enteric log log}\frac{CFU}{mL}\textit{on the enteric at 0 h}\right)}{\left(\textit{enteric log log}\frac{CFU}{mL}\textit{on glucose at 48 h} - \textit{enteric log log}\frac{CFU}{mL}\textit{on glucose at 0 h}\right)}\right] \quad (1)$$

And the metabolism can be assessed by changes in pH and short-chain fatty acids (SCFA) production [16].

3.2.3 Short-Chain Fatty Acids Quantification

The SCFA quantification can be accessed by chromatographic analysis as proposed by Zhao et al. [14] and Silva et al. [17].

1. Prepare standard solutions of acetic acid (0.04–10.49 mg mL^{-1}), propionic acid (0.04–9.93 mg mL^{-1}) and butyric acid (0.08–9.64 mg mL^{-1}).
2. Prepare a 2-ethylbutyric acid solution containing HCl 5 mol L^{-1} used as internal standard stock solution.
3. Centrifuge the samples (10 min, 10,000 RPM and 5°C).
4. Collect the supernatant and filter using 0.22 μm filters.
5. Adjusted of the pH to 2–3 by adding acidified water.
6. Let the room temperature rest for 10 min with occasional shaking.
7. Immediately before the analysis, add to the supernatant the internal standard 2-ethylbutyric acid solution up to the final concentration of 1 mmol L^{-1} and inject the solution into the gas chromatographic system.
8. For the chromatographic analysis, use a system equipped with a flame ionization detector (FID) and a capillary column (30 m × 0.25 mm × 0.25 μm).
9. Use the following conditions: injector and detector temperatures set at 250°C, volume of injection of 1 μL with the split ratio set to 1:10, and helium as carrier gas at 1.0 mL min^{-1}.

10. The oven conditions should be: initial temperature of 100°C, maintained for 0.5 min, raised to 180°C at 8°C min^{-1}. It must be maintained for 1.0 min and then increased to 200°C at 20°C min^{-1}. Finally held at 200°C for 5 min in a total of 17.5 min.
11. The composition of SCFAs will be identified by their retention times compared with the standard solution.

3.3 In Vivo Study—Animals

It is possible to investigate in a short or long-term feeding different resistant starch diets exhibit prebiotic effects by altering the gut bacterial abundance and promoting its metabolic activity using the rat as a model of mammalian digestive physiology [18].

3.3.1 Animals

1. Conduct the experiment with male or female Wistar rats weaned at three weeks of age, selecting the number of each experimental group.
2. Place the animals in individual cages under controlled environmental conditions in order to maintain a temperature of 22 ± 2°C, a relative humidity of 50%–60% and an light/dark cycle of 12 h, ad libitum access to water and food during 1 week for acclimatization [18, 19].

3.3.2 Diet

1. Designate randomly the animals in experimental diet groups, either in the presence or in absence (control groups) of resistant starch addition test formulations.
2. Feed the animals with treatment diets for four weeks, given ad libitum access to their allocated experimental diet and water during the experiment.
3. Measure food and water intake every day during the experiment, and weigh once a week to monitor weight gain [2].
4. At the end of the study anesthetized the rats with via an intramuscular injection with xylazine (10 mg/kg) and ketamine (150 mg/kg) following sacrifice [20].

3.3.3 DNA Extraction and Microbial Profiling

1. To assess the impact of the diet on the animals' intestinal microbiota, perform feces collection on days 0, 1, 14, and 28. The samples on day 0 will represent the baseline, with all groups previously fed with the control diet.
2. Extract genomic DNA from fecal samples of animals using the QIAmp DNA Stool kit according to the manufacturer's protocol.
3. Amplify the V4–V6 variable regions of the 16S rRNA gene by PCR and then sequence using the following primers: F515 (5′GTGCCAGCMGCCGCGGTAA-3′) and RNextera (5′ CGACRRCCATGCANCACCT-3′)according to Klingbeil

et al. [21] following the instructions below, with some modifications.

4. Perform the sequencing on an Illumina MiSeq platform.
5. For the preparation of the libraries, use the kit "Illumina TruSeq DNA Sample Preparation v2."
6. Execute quality control of the initial sequences through processing data, applying FASTQC, and then performing the quality filtering using the Trimmomatic tool (v 0.36) [21].
7. Analyze the sequences using the software Quantitative Insights into Microbial Ecology (QIIME). In this step, execute a chimera search using the *usearch* method.
8. Cluster high-quality reads into operational taxonomic units (OTUs) at a 97% similarity level.
9. Access alpha diversity indices (Chao1, ACE, Shannon and Simpson) and perform beta diversity through the principal component analysis (PCA) and/or heat map to assess patterns in the data sets (for example, abundance of genera/species × treatments), using the XLSTAT software as describe by Shang et al. [22] with some modifications.

3.3.4 Short-Chain Fatty Acid (SCFA) Analysis

Total acetate, propionate, butyrate and total short-chain fatty acids (SCFAs) should be extracted, and the concentrations should be determined by gas chromatography (GC) according to the item (3.2.3).

3.3.5 Evaluation of Animal Health Parameters

1. After 12 h of fasting, the animals in the studied groups will be weighed and then subjected to a fasting blood glucose test through the caudal cava vein (time 0), with the use of a glucometer.
2. Apply to the animals an intraperitoneal injection of glucose at a concentration of 2 g/kg body weight and verify the blood glucose at 15, 30, 60, and 120 min after glucose application.
3. In order to perform the insulin tolerance test, animals in the fed state must be weighed and checked for blood glucose in the fed state (time 0). Then, apply 1.5 U/kg of insulin via intraperitoneal, and verify blood glucose at 3, 6, 9, 12, and 15 min.
4. Collect blood samples through portal vein puncture using a syringe and centrifuged 3000 × *g* for 15 min at 4°C to obtain the serum [20].
5. Determine serum dosages of uric acid, creatinine, AST (aspartate aminotransferase), ALT (alanine aminotransferase), blood glucose, total proteins, albumin, triacylglycerol (TG), total cholesterol (TCH), high-density lipoprotein cholesterol (HDL-c), and low-density lipoprotein cholesterol (LDL-c) using commercial kits according to the manufacturer's instructions.

3.4 In Vivo Study—Human

Resistant starch can be determined as a prebiotic tool for health by verifying its interaction with the intestinal microbiota in studies involving human volunteers.

3.4.1 Experimental Design—Selection of Healthy Individuals

The parameters for selecting individuals to conduct the experiments with the consumption of a diet with resistant starch supplementation may diverge according to the study group of interest. For example, if the purpose is to access the prebiotic potential of starch in healthy humans, the selection and exclusion criteria of individuals for that study will be distinct when compared to other groups that have specific metabolic diseases, such as obesity, type 2 diabetes mellitus and others.

1. Establish age range of eligible individuals (under the age of 18 with permission from a parent or legal guardian, or up to 18 years or older);
2. Establish a weight standard by waist circumference (<85 cm in men and <80 cm in women), and through the body mass index of healthy individuals (18.5–24 kg/m^2).

3.4.2 Exclusion Criteria in Healthy Individuals

1. Presence of metabolic or chronic diseases; supplementation of probiotics or use of antibiotics in the last three months before the study; participation in other clinical trials in the four weeks prior to the study; use of medications that can control blood glucose, blood cholesterol and/or blood pressure; smoker; suffered from gastrointestinal problems such as irritable bowel syndrome, inflammatory bowel disease, or colorectal cancer [23–27]. If the individuals fall into any parameter present in the exclusion criteria, they should be discontinued from the study.

If the aim of the study involves determining the effects of resistant starch on metabolic parameters and inflammation in individuals classified with metabolic disorders, different selection criteria need to be adopted, necessitating a revision of the weight standard, body mass index, and exclusion criteria.

3.4.3 Diet

1. Perform the diet according to the usual consumption of each individual. Likewise, maintain their current physical activities [28, 29].
2. Collect information about the dietary intake of macronutrients and analyze it at the beginning and at the end of the study. Study participants should receive a diary and be oriented to record daily information regarding their diet in order to assess food intake and intestinal habits in each treatment period. Likewise, they should report any possible changes that have occurred, such as changes in medications or changes in usual exercise habits and other relevant information.

3. Determine, according to the application study, the dosage of resistant starch and its form of administration, as well as the period(s) of the day that will be consumed.
4. The safe consumption of resistant starch should be evaluated. For this, it is necessary that individuals complete a symptom diary to assess the potential side effects of RS administration. The symptoms include bowel movement, stool consistency, discomfort, flatulence, abdominal pain, and bloating, and subjects should be asked to score them on a scale from 1 (none, normal, good well-being) to 5 (severe symptoms and discomfort) [23].

3.4.4 Conduct of the Study

1. Calculate de minimum sample size of individuals for the study based on Pocock's formula [30].
2. The study should be conducted in a randomized, double blind, crossover and placebo-controlled manner.
3. Randomly assigned the participants to study groups and divided the samples containing resistant starch and placebo among volunteers based on allocation codes after randomization. Therefore, researchers analyzing samples have to always be blinded to the composition associated with the samples as well as the statistician until the end of the analysis.
4. Start diet administration with a 2-week baseline diet (without resistant starch administered), followed by administration of resistant starch and placebo samples, each one offered over a 2-week period interspersed by 2-week washout periods (baseline diet), and a final washout period.

3.4.5 Fecal Sample Collection and 16S rRNA Gene Sequencing

When the resistant starch present in the diet reaches the colon in its accessible form for microbial digestion, it is possible to study and determine the modulation of specific members of the gut microbiome. Hence, analysis of stool samples and 16S rRNA gene sequencing are aimed at understanding the impacts of RS at the microbial community level, allowing for the observation of responses to prebiotic performance. The methodology can be performed as proposed by Venkataraman et al. [26] Maier et al. [25] and Baxter et al. [31, 32].

1. Perform the collection of fecal samples before and after each intervention regarding baseline diet and each diet with either RS or placebo administration weekly in collection tubes using an OMNIgene-Gut (DNA Genotek) collection kit, adding the stabilizing buffer solution and homogenizing the stool samples following the process according to the manufacturer's protocol.
2. Store the collection tubes containing the fecal samples within 24 h of collection in a −20°C freezer until use for DNA and

metabolite extractions. Weigh the collection tubes before and after fecal collection to determine the weight of the fecal material collected.

3. Extract DNA from stool samples in duplicate using 0.25 mL from fecal suspension using a MoBioPowerMag Soil DNA kit following the manufacturer's instructions.
4. Execute the amplification of the V4 hypervariable region of the 16S rRNA from each sample through PCR using 2×250-bp paired-end kits, and perform sequencing on an Illumina MiSeq platform.
5. For the preparation of the libraries, use the kit "Illumina TruSeq DNA Sample Preparation v2."
6. Execute quality control of the initial sequences through processing data, applying FASTQC and then performing the quality filtering using the Trimmomatic tool (v 0.36) [32, 22].
7. Analyze the sequences using the software Quantitative Insights into Microbial Ecology (QIIME). In this step, execute a chimera search using the *usearch* method.
8. To access the taxonomic profiles of the samples, clustering the sequences into operational taxonomic units (OTUs) using a closed-reference OTU picking protocol at 97% sequencing identity aligned to the SILVA reference database.

3.4.6 Short-Chain Fatty Acid (SCFA) Analysis

In order to evaluate the impact of resistant starch samples as a prebiotic agent through the concentration of SCFAs in the individual, it is possible to analyze stool samples, as described in item (3.2.3).

3.4.7 Biochemical Analyzes

Study participants shall attend the research center to perform 12 h fasting blood biochemical tests obtained in the morning prior to commencing each phase and at the end during the final week of each dietary intervention. The analysis can be performed according to Gargari et al. [4] and Bodinham et al. [9].

1. After overnight fasting, collect the venous blood from the volunteers and transfer it to Vacutainer tubes.
2. After collection, centrifuge blood samples at 1048 × g at room temperature, separating serum from whole blood and proceed with all the analysis in duplicate [4].
3. Measure serum insulin concentration with ELISA (human insulin ELISA kit) with inter-assay coefficient of variability (CV) < 14% and intra-assay (CV) < 10% [9].
4. Calculate HOMA-IR from serum insulin and glucose concentrations as described below in Eq. 2:

$$\text{HOMA} - \text{IR} = \left[insulin \left(\frac{\propto U}{L} \right) \times \frac{glucose \left(\frac{nmol}{L} \right)}{22.5} \right] \tag{2}$$

5. Perform analysis of serum levels of total cholesterol (TC), triglycerides (TG), high-density lipoprotein cholesterol (HDL-c), and glucose concentrations using commercial kits. Calculate low-density lipoprotein cholesterol (LDL-c) using the Friedewald equation described below in Eq. 3:

$$\text{LDL cholesterol} = [\text{Total cholesterol}] - [\text{HDL cholesterol}] - \left[\frac{\text{Triglyceride}}{5} \right] \tag{3}$$

6. For the measurements of the following inflammatory biomarkers, tumor necrosis factor α (TNFα) and interleukin 6 (IL6), apply commercially available ELISA kits, with inter-assay coefficient of variability (CV) values <4 and <18%, respectively and intra-assay (CV) values <7 and <10%, respectively.

4 Notes

4.1 MRS Broth Preparation

The modified MRS broth can be prepared according de Albuquerque et al. [14] using:tryptone 10 g L^{-1}, meat extract 8 g L^{-1}, yeast extract 4 g L^{-1}, di-potassium hydrogen phosphate 2 g L^{-1}, tween 80 1 g L^{-1}, sodium acetate 5 g L^{-1}, tribasic ammonium citrate 2 g L^{-1}, magnesium sulfate 0.2 g L^{-1}, manganese sulfate 0.04 g L^{-1}, and the respective sugar source 20 g L^{-1} (glucose, lactose, and fructooligosaccharide).

4.2 Seeds

First, fix the starch with a disodium citric acid-hydrogen phosphate buffer solution, followed by the addition of a resistant α-amylase at a temperature of 90°C in an orbital incubator shaker at 128 rpm for 2 h. Then, the pH should be adjusted to 4.5 with a 4 mol L^{-1} citric acid solution. Add the enzyme glucoamylase (300 U mL^{-1}) to a concentration of 5000 U g^{-1} of starch, followed by incubation at 60°C for 1 h in the orbital incubator shaker at 128 rpm. The samples must be centrifuged at 3000*g* for 15 min, and the supernatant must be discarded. The resulting precipitate must be washed with distilled water, and the process must be repeated two more times. The precipitate must be dried in an oven at 50°C, crushed and sieved through a 185 μm mesh screen [1, 33].

4.3 Fruits

The fruits must be washed, peeled, and the seeds removed. The pulp must be cut into small pieces and placed in a 200 mg L^{-1} solution of sodium bisulfite. After that, the fruits must be washed in water and crushed in a liquid with 1:5 (v/v) distilled water until a homogeneous suspension is formed. The suspension must be sieved through a 22 mesh sieve. The retained portion must be discarded and the permeate must be refrigerated over night to decant the starch. The starch must then be dried at 50°C for approximately 15 h in a dry hot air convection dryer [13, 34].

Acknowledgments

This work was supported Coordenação de Aperfeiçoamento de Pessoal de Nível Superior—Brasil (CAPES)—Finance code 001.

References

1. Su H, Chen J, Miao S, Deng K, Liu J, Zeng S et al (2019) Lotus seed oligosaccharides at various dosages with prebiotic activity regulate gut microbiota and relieve constipation in mice. Food Chem Toxicol 134:110838
2. Shang W, Si X, Zhou Z, Wang J, Strappe P, Blanchard C (2017) Studies on the unique properties of resistant starch and chito-oligosaccharide complexes for reducing high-fat diet-induced obesity and dyslipidemia in rats. J Funct Foods 38:20–27
3. Lu X, Chen J, Guo Z, Zheng Y, Rea MC, Su H et al (2017) Using polysaccharides for the enhancement of functionality of foods: a review. Trends Food Sci Technol 86:311–327
4. Gargari BP, Namazi N, Khalili M, Sarmadi B, Jafarabadi MA, Dehghan P (2015) Is there any place for resistant starch, as alimentary prebiotic, for patients with type 2 diabetes? Complement Ther Med 23(6):810–815
5. Topping DL, Clifton PM (2001) Short-chain fatty acids and human colonic function: roles of resistant starch and nonstarch polysaccharides. Physiol Rev 81:1031–1064. p
6. Cerqueira FM, Photenhauer AL, Pollet RM, Brown HA, Koropatkin NM (2020) Starch Digestion by gut bacteria: crowdsourcing for carbs. Trends Microbiol 28(2):95–108
7. Bindels LB, Walter J, Ramer-Tait AE (2015) Resistant starches for the management of metabolic diseases. Curr Opin Clin Nutr Metab Care 18(6):559–565
8. Ze X, Duncan SH, Louis P, Flint HJ (2012) *Ruminococcus bromii* is a keystone species for the degradation of resistant starch in the human colon. ISME J 6(8):1535–1543
9. Bodinham CL, Smith L, Thomas EL, Bell JD, Swann JR, Costabile A et al (2014) Efficacy of increased resistant starch consumption in human type 2 diabetes. Endocr Connect 3(2):75–84
10. Karimi P, Farhangi MA, Sarmadi B, Gargari BP, Zare Javid A, Pouraghaei M et al (2021) The therapeutic potential of resistant starch in modulation of insulin resistance, endotoxemia, oxidative stress and antioxidant biomarkers in women with type 2 diabetes: a randomized controlled clinical trial. Ann Nutr Metab 68(2):85–93
11. Snelson M, Jong J, Manolas D, Kok S, Louise A, Stern R et al (2019) Metabolic effects of resistant starch type 2: a systematic literature review and meta-analysis of randomized controlled trials. Nutrients 11:1833. p
12. Goñi I, García-Diz L, Mañas E, Saura-Calixto F (1996) Analysis of resistant starch: a method for foods and food products. Food Chem 56(4):445–449
13. Pereira APA, Lauretti LBC, Alvarenga VO, Paulino BN, Angolini CFF, Neri-Numa IA et al (2020) Evaluation of fruta-do-lobo (*Solanum lycocarpum* St. Hill) starch on the growth of probiotic strains. Food Res Int 133:109187
14. De Albuquerque TMR, Borges CWP, Cavalcanti MT, Lima M, dos S, Magnani M, de Souza EL (2020) Potential prebiotic properties of flours from different varieties of sweet potato (*Ipomoea batatas* L.) roots cultivated in Northeastern Brazil. Food Biosci 36:100614
15. Zhang S, Hu H, Wang L, Liu F, Pan S (2018) Preparation and prebiotic potential of pectin oligosaccharides obtained from citrus peel pectin. Food Chem 244:232–237
16. Zhao G, Nyman M, Jönsson JÅ (2006) Rapid determination of short-chain fatty acids in colonic contents and faeces of humans and rats by acidified water-extraction and

direct-injection gas chromatography. Biomed Chromatogr 20(8):674–682
17. da Silva JK, Cazarin CBB, Bogusz Junior S, Augusto F, Maróstica Junior MR (2014) Passion fruit (*Passiflora edulis*) peel increases colonic production of short-chain fatty acids in Wistar rats. LWT Food Sci Technol 59(2):1252–1257
18. Paturi G, Butts CA, Stoklosinski H, Herath TD, Monro JA (2017) Short-term feeding of fermentable dietary fibres influences the gut microbiota composition and metabolic activity in rats. Int J Food Sci Technol 52(12):2572–2581
19. Krupa-Kozak U, Juskiewicz J, Wronkowska M, Soral-Śmietana M, Zduńczyk Z (2010) Native and microwaved bean and pea starch preparations: physiological effects on the intestinal ecosystem, caecal tissue and serum lipids in rats. Br J Nutr 103(8):1118–1126
20. Roquetto AR, Monteiro NES, Moura CS, Toreti VC, de Pace F, Santos Ados et al (2015) Green propolis modulates gut microbiota, reduces endotoxemia and expression of TLR4 pathway in mice fed a high-fat diet. Food Res Int 1(76):796–803
21. Klingbeil EA, Cawthon C, Kirkland R, de La Serre CB (2019) Potato-Resistant starch supplementation improves microbiota dysbiosis, inflammation, and gut–brain signaling in high fat-fed rats. Nutrients 11(11):2710
22. Bolger AM, Lohse M, Usadel B (2014) Trimmomatic: a flexible trimmer for Illumina sequence data. Bioinformatics 30(15):2114–2120
23. Shang W, Si X, Zhou Z, Li Y, Strappe P, Blanchard C (2017) Characterization of fecal fat composition and gut derived fecal microbiota in high-fat diet fed rats following intervention with chito-oligosaccharide and resistant starch complexes. Food Funct 8(12):4374–83.
24. Martínez I, Kim J, Duffy PR, Schlegel VL, Walter J (2010) Resistant starches types 2 and 4 have differential effects on the composition of the fecal microbiota in human subjects. PLoS ONE 5(11):e15046
25. Zhang L, Ouyang Y, Li H, Shen L, Ni Y, Fang Q et al (2019) Metabolic phenotypes and the gut microbiota in response to dietary resistant starch type 2 in normal-weight subjects: a randomized crossover trial. Sci Rep 9(1):1–11
26. Maier TV, Lucio M, Lee LH, Verberkmoes NC, Brislawn CJ, Bernhardt J et al (2017) Impact of dietary resistant starch on the human gut microbiome, metaproteome, and metabolome. mBio 8(5):1343–1360
27. Venkataraman A, Sieber JR, Schmidt AW, Waldron C, Theis KR, Schmidt TM (2016) Variable responses of human microbiomes to dietary supplementation with resistant starch. Microbiome 4(1):1–9
28. Stewart ML, Zimmer JP (2018) Postprandial glucose and insulin response to a high-fiber muffin top containing resistant starch type 4 in healthy adults: a double-blind, randomized, controlled trial. Nutrition 1(53):59–63
29. Venkataraman A, Sieber JR, Schmidt AW, Waldron C, Theis KR, Schmidt TM (2016) Variable responses of human microbiomes to dietary supplementation with resistant starch. Microbiome 4:1–9
30. Stewart ML, Nikhanj SD, Timm DA, Thomas W, Slavin JL (2010) Evaluation of the effect of four fibers on laxation, gastrointestinal tolerance and serum markers in healthy humans. Ann Nutr Metab 56(2):91–98
31. Pocock SJ (2013) Clinical trials: a practical approach. John Wiley & Sons
32. Baxter NT, Schmidt AW, Venkataraman A, Kim KS, Waldron C, Schmidt TM (2019) Dynamics of human gut microbiota and short-chain fatty acids in response to dietary interventions with three fermentable fibers. mBio 10(1):e02566-18. https://doi.org/10.1128/mBio.02566-18
33. Zeng S, Wu X, Lin S, Zeng H, Lu X, Zhang Y et al (2015) Structural characteristics and physicochemical properties of lotus seed resistant starch prepared by different methods. Food Chem 186:213–222
34. Clerici MTPS, Kallmann C, Gaspi FOG, Morgano MA, Martinez-Bustos F, Chang YK (2011) Physical, chemical and technological characteristics of *Solanum lycocarpum* A. St. - HILL (Solanaceae) fruit flour and starch. Food Res Int 44(7):2143–2150

Check for updates

Chapter 20

Water Absorption Index and Water Solubility Index

Hugo José Martins Carvalho, Bruna Lago Tagliapietra, Natali Alcantara Brandão and Maria Teresa Pedrosa Silva Clerici

Abstract

The water absorption index (WAI) and water solubility index (WSI) are parameters used to evaluate the functional behavior of native or modified starches. WAI measures the starch granules' ability to absorb and retain water, while WSI reflects the release of soluble components such as glucose, maltose, and low-molecular-mass dextrins, as a result of physical and/or chemical starch modification treatments. For example, pre-gelatinized starches obtained by hot extrusion are evaluated for their WAI and WSI content to determine as processing conditions. These analyses aid in evaluating the technological performance of starches and their potential industrial applications, particularly as thickening agents in soups, creams, broths, and related products. This chapter also describes in detail the standardized experimental protocol for determining both indices.

Keywords Swelling, Hydration, Starch, Gelatinization, Dextrinization

1 Introduction

Assessing the technological characteristics of starch is an important step in identifying potential applications and determining whether modifications have occurred that altered its physical properties.

The water absorption index (WAI) and water solubility index (WSI) methods are considered rapid methods, without the use of chemical reagents and with easy interpretation of results. When performed together, both processes can provide a comparative analysis of which processes had the greatest impact on starch. Starch is considered insoluble and has no water absorption capacity in granular form. It has a density ($d_{\text{corn starch}} = 1.5$ g/mL) greater than that of water ($d_{\text{water}} = 1$ g/mL). Therefore, if the starch is damaged or gelatinized, it does not completely separate from the water. It can absorb this water if partially or fully gelatinized and/or be partially or fully solubilized, depending on how much it has been transformed into low molecular weight dextrins, maltose, and glucose.

Maria Teresa Pedrosa Silva Clerici and Pedro H. Campelo (eds.), *Basic Protocols in Starch Characterization*, Methods and Protocols in Food Science, https://doi.org/10.1007/978-1-0716-5116-2_20,

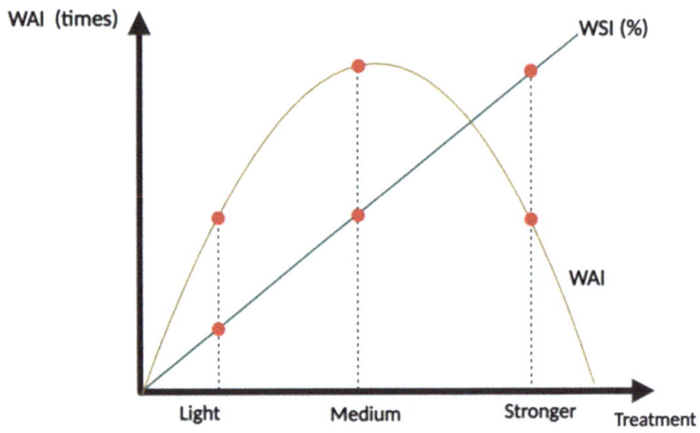

Fig. 1 Effect of thermomechanical treatment on starch modification. Source: Authors' collection, 2025

WAI and WSI data should not be analyzed separately. If possible, the original raw starch should be used when analyzing the effects of starch extraction processes or physical, enzymatic, chemical, or genetic modifications. For example, Fig. 1 illustrates the effects of these indices during thermomechanical starch processing. In a gentle treatment process, both WAI and WSI are low. As the treatment becomes more intense, WAI tends to increase, as does WSI. However, when the treatment is very intense, for example, under high temperature and pressure conditions, WAI decreases because the starch becomes extremely dextrinized, and WSI increases. Therefore, to comparatively identify the effect of processing on these two indices, it is important to perform both indices together and also have a control sample, preferably of the original native starch.

2 Material

2.1 Equipment

1. Analytical balance with precision up to 0.0001 g, within calibration period and with available standard operating procedure.
2. Drying oven with air circulation.
3. Rotating centrifuge for 50 mL Falcon-type tubes with a cooling system.
4. Desiccator.

2.2 Material

1. Disposable gloves.
2. Metal spatula.

3. 50 mL Falcon tubes with caps, previously weighed and labeled (tube and cap).
4. Distilled water wash bottle.
5. Petri dishes.
6. 50 mL graduated cylinder.
7. Spatula.

2.3 Reagents

1. Freshly distilled deionized water.

3 Method

Determination of the absorption index and solubility in water, based on the methodology of Anderson [1], with no WSI, because to Anderson, the supernatant liquid was poured carefully into a tared evaporating dish and the index of water solubility was the amount of dried solids recovered by evaporating the supernatant from the water absorption test was expressed as percentage of dry solids in the 2.5 g. After several experiments, we have seen that comparative process studies can alter supernatant volumes and lead to calculation errors. Therefore, we prefer to demonstrate how to remove an aliquot of the supernatant from a previously weighed plate, evaporate the water, and weigh the dehydrated soluble solid. Calculations are based on the total weight of the sample and water. Soluble solids, such as glucose and dextrins, are distributed throughout the sample and will not be separated by centrifugation to the point where they remain only in the supernatant. Examples of this calculation will be shown at the end of this chapter.

3.1 Sample

Dry starch with previously determined moisture content (Chapter 1 of this book).

3.2 Procedure

3.2.1 Preanalysis Steps

1. Separate the number of tubes and caps according to the number of samples and perform the analyses in triplicate. Check that they are prelabeled and resistant to washing and heating. If not, use a high-temperature pen to identify them.
2. Separate the number of Petri dishes according to the number of samples and perform the analyses in triplicate. Dry the plates in an air-circulating oven at 130°C for 1 h. After the stipulated time, remove the plates from the oven and place them in a desiccator for 30 min (or until cool, but never exceeding 40 min). Weigh the plates and keep them in a desiccator until the moment of use.

3. Check the operating procedure of the air circulation oven and whether it is precalibrated. Record this information. Strictly follow the operating procedure.
4. Check the standard operating procedure of the analytical balance and whether it is calibrated. Record the calibration date. Strictly follow the operating procedure.
5. To collect data, prepare a results table in the laboratory notebook, as shown in Tables 1 and 2 in **Results**.

3.2.2 Water Absorption Index (WAI) and Water Solubility Index (WSI)

1. Identify the Falcon tubes and their respective caps.
2. Weigh the tubes (tube + cap) on an analytical balance (0.001 g accuracy) and record the exact weight.*
   *Perform the process in triplicate.
3. Add 2.5 g of the sample and record the exact weight as shown in Fig. 2A.
4. Add 30 mL of distilled water to the tube and weigh again (P_{total}).
5. Perform intermittent agitation for 30 min, as shown in Fig. 2(B).*
   **Intermittent agitation consists of shaking the tube containing the sample and water for 5 min, followed by a 5-min rest period, repeating the process until the stipulated time (30 min) is reached.*
6. Transport the tubes containing the starch to the centrifuge (*see* **Note 1**).*
   **Follow centrifuge calibration procedures as described in Supplementary Material 2 in* Chapter 6.
7. Centrifuge at room temperature (~25°C) at 3000 rpm for approximately 10 min.*
8. Pipette 5 ml of the supernatant from the tube into previously weighed Petri dishes (if there is insufficient supernatant, reduce the sample mass, or change the proportion between sample and water, to increase the new value of water) and proceed to determine the WSI, as shown in Fig. 2c and continue in step 11.
9. Discard the remaining soluble fraction in the beaker after step 8.
10. Weigh the tube with centrifugation residue on an analytical balance (accuracy of 0.001 g) and record the exact weight ($P_{tube+residue}$).

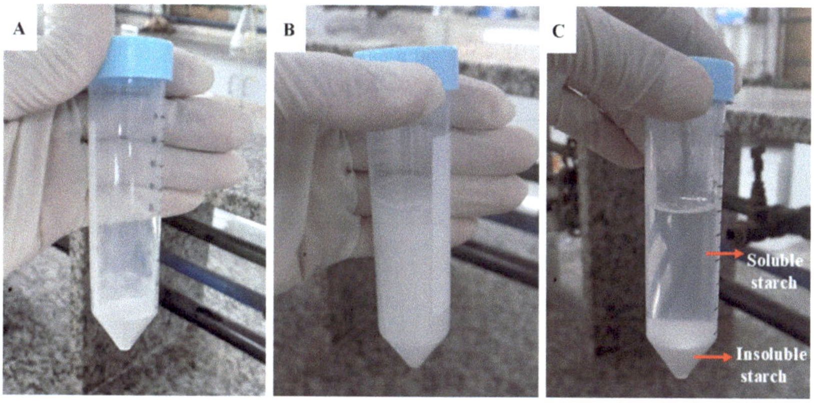

Fig. 2 Weighing (**A**), mixing/homogenization (**B**) and obtaining the soluble fraction and residue (**C**) steps to determine the water solubility index and water absorption index for starchesSource: Authors' collection, 2025

11. Dry the plates in an air-circulating oven at an initial temperature of 50°C for 2 h or until the water evaporates slowly from the plate (about 3–4 h, depending on the oven and quantity of plates), then increase the temperature to 105°C for 4 h to determine the soluble supernatant residue.
12. After the stipulated time, remove the plates from the oven and place them in a desiccator for 30 min (or until cool, but never exceeding 40 min) (P_{soluble}).
13. After the stipulated time, weigh the plates and record the weight.

3.3 Results

Perform the calculation for the water absorption index according to the following equation:

$$\text{WAI} = \frac{P_f}{P_a}$$

Where

$$P_f = \text{weight of centrifugation residue (g)} = \text{inicial weigth} - \text{final weight}$$

$$P_a = \text{initial sample weight on a dry basis}$$

Perform the calculation for the water solubility index according to the following equation:

$$P_{\text{soluble}} \text{———————————} P_{\text{plate}}$$

$$x \text{———————————} P_{\text{total}}$$

$$x = \frac{P_{\text{soluble}} \times P_{\text{total}}}{P_{\text{plate}}}$$

$$\text{WSI}(\%) = x \times 100$$

Where

$$P_{\text{soluble}} = \text{ weight of soluble residue in the plate (g)}$$

$$P_{\text{plate}} = \text{ initial plate weight}$$

$$P_{\text{total}} = \text{ initial sample weight} + 30 \text{ mL water}$$

Multiplication by 6 is based on removing only 1/6 of the water fraction of item 2.5.3. (1), hence the need for multiplication to reflect the real value of the soluble fraction (*see* **Note 2**).

After obtaining the yield of the three replicates per sample, calculate the mean, standard deviation (SD), and coefficient of variation (CV) according to equations provided by data tabulation programs, such as Microsoft Excel or Google Sheets.

The standard deviation is considered optimal when its value is as close to zero as possible. A coefficient of variation below 10% indicates a dispersion close to ideal for this type of analysis.

Table 1
Example of data collected in the analysis

Determination of water absorption index (WAI)										
Sample	**Cassava starch—CS**									
Analyst	**Hugo**									
Date	**20/07/2025**									
Sample identification (code and reference)	**Number**	**Replica**	**Weight of tube + cap (g) (P1)**	**Initial sample weight (g) (P2)**	**Weighing tube weight + starch residue (g) (P3)**	**Final starch weight (g) P3 - P1**	**WAI**	**Average**	**DP**	**CV**
	1	CS1	10.2536	2.5423	20.9789	10.7253	4.22			
CS*	2	CS2	11.4987	2.5214	21.9256	10.4269	4.15	4.24	0.12	2.82
	3	CS3	10.9536	2.5863	22.2598	11.3062	4.37			

*Example of data recording for a starch sample. *CS*, cassava starch; *SD*, standard deviation; *CV*, coefficient of variation

Table 2

Example of data collected in the analysis

Determination of water solubility index										
Sample	**Cassava starch**									
Analyst	**Martins**									
Date	**20/07/2025**									
Sample identification (code and reference)	**Number**	**Replica**	**Empty plate weight (g) (P1)**	**Initial sample weight (g) (P2)**	**Weighing empty plate + dry starch (g) (P3)**	**$P_{soluble}$ (g) (P3 - P1)**	**WSI (%)**	**Average (%)**	**DP**	**CV**
	1	CS1	23.5452	2.5423	23.6011	0.0558	13.1810			
CS*	2	CS2	22.2235	2.5214	22.2698	0.0463	11.0256	12.1661	0.8844	0.0727
	3	CS3	21.7963	2.5863	21.8493	0.529	12.2917			

*Example of data recording for a starch sample. *CS*, cassava starch; *SD*, standard deviation; *CV*, coefficient of variation.

4 Notes

1. Fill all capped tubes with the same sample level. Furthermore, the tubes should be placed two at a time (always in pairs) on opposite sides of each other to prevent the centrifuge from becoming unbalanced and potentially stalling.
2. The difference in units of expression between the WAI and the WSI stems from the nature of the measurements each represents. The WAI quantifies how much water is retained per gram of sample and is not expressed as a percentage because it is a ratio of absolute mass (mass of water absorbed/mass of sample), so expressing it as a percentage would be meaningless, as it is not a fraction of the whole. The WSI quantifies the fraction of the mass of the sample solubilized in water relative to the total mass, and is expressed as a percentage, because the values are very low.

Acknowledgments

The authors would like to thank the National Council for Scientific and Technological Development (CNPq), Brazil, for the scholarships awarded to MTP Clerici (Processes 312660/2023-5). Additionally, the authors express gratitude to the Coordination for the Improvement of Higher Education Personnel (CAPES, financial code 001), Brazil, for the scholarships awarded to NA Brandão (Process 88887.605710/2021-00), HJMC (Process 88887.141456/2025-00) and for supporting the Graduate Program in Food Technology at the Universidade Estadual de Campinas (UNICAMP), São Paulo, Brazil, and the São Paulo Research Foundation (FAPESP), Brazil (Process Number 2023/12726-8) for the postdoctoral fellowship to BL Tagliapietra.

Reference

1. Anderson, R. A. (1982). Water absorption and solubility and amylograph characteristics of roll-cooked small grain products. Cereal Chemistry 59(4):265–269

Check for updates

Chapter 21

Oil Holding Capacity

Bruna Lago Tagliapietra, Hugo José Martins Carvalho, Natali Alcantara Brandão and Maria Teresa Pedrosa Silva Clerici

Abstract

Oil holding capacity (OHC) is an essential functional property of starches and vegetable flours. This chapter presents a standardized method for determining the OHC of starch samples, using a cold emulsification process followed by centrifugation. The methodology enables quantification of the amount of oil retained within the sample matrix through gravimetric analysis. Experimental procedures, equipment, and technical precautions are detailed to ensure the reproducibility and reliability of results. The data obtained are essential for identifying new starch sources with high OHC, adding value to underexplored botanical sources.

Keywords Starch, Modified starches, Fat

1 Introduction

Oil retention capacity (ORC) is an important functional property of food ingredients, especially vegetable flours and starch, which directly influences the texture, palatability, and stability of emulsions in processed products. This property is related to the sample's ability to absorb and retain lipids in its structural matrix and is essential in the formulation of foods such as sausages, baked goods, and meat and dairy products, such as reduced-fat cream cheeses, dairy desserts, salad dressings, mayonnaise, and others [1].

The growing search for fat and oil substitutes has led to advances in the field of modified starches. These starches, in their native form, have low oil absorption capacity, but when modified, they exhibit oil-absorbing properties, which lead to greater stability in formulations [2].

The most commonly used modified starches for oil absorption are chemically modified, such as esterified, etherified, phosphated, and oxidized starches, among others. However, physical modifications have also become more prominent, such as the use of plasma to

Maria Teresa Pedrosa Silva Clerici and Pedro H. Campelo (eds.), *Basic Protocols in Starch Characterization*, Methods and Protocols in Food Science, https://doi.org/10.1007/978-1-0716-5116-2_21,

increase pores, micronization of starches, making them more brittle and porous [3].

It is worth noting that the starches being presented as new potential sources to replace corn, rice, cassava, potato, and wheat starches, which are currently commercially available, may have different characteristics. Therefore, studying this property will be important, as if it proves positive, it could add greater added value to the starch, as it could be used as a native starch in formulations.

2 Material

2.1 Equipment

1. Analytical balance with precision up to 0.0001 g, within calibration period and with available standard operating procedure.
2. Vortex mixer.
3. Rotary centrifuge for 50 mL Falcon tubes with a temperature control system.
4. Stopwatch.

2.2 Materials

1. Disposable gloves.
2. Starch sample with previously determined moisture content.
3. Paper towels or tissue for cleaning the outside of the tubes, if necessary.
4. 50 mL Falcon tubes with caps.
5. Racks for 50 mL Falcon tubes.
6. Glass beakers.
7. Metal spatula.

2.3 Reagents

1. Soy oil.

3 Method

The methodology described below allows quantifying the CRO based on the difference in mass between the added oil and the mass of oil not retained by the sample analyzed after centrifugation [1, 3, 4].

3.1 Sample

Dry starch with previously determined moisture content (Chapter 1 of this book).

3.2 Procedure

This protocol aims to determine the oil holding capacity under low temperature conditions (cold emulsification), using a method adapted by Lin et al. (1974).

3.2.1 Preanalysis Steps

1. Separate the number of tubes according to the number of samples and perform the analyses in triplicate. Check that they are prelabeled and resistant to washing and heating. If not, use a high-temperature pen to identify them.
2. Check the standard operating procedure of the analytical balance and whether it is calibrated. Record the calibration date. Strictly follow the operating procedure.
3. To collect data, prepare a results table in the laboratory notebook, as shown in Table 1.

3.2.2 Oil Holding Capacity

1. Identify the Falcon tubes and their respective caps.
2. Weigh the tubes (tube + cap) on an analytical balance (0.001 g accuracy) and record the exact weight, record the value as P_1.*

   **Perform the process in triplicate.*
3. Add 1.5000 g of the dry starch sample directly to the tube and weigh again.
4. Record the weight of the tube with the sample (P_2).
5. Add 15 mL of soybean oil (at 25°C) to the tube, mix gently, and weigh the whole (tube with the cap + sample + oil). Record as P_3.
6. Weigh a clean, dry beaker (previously weighed) and record the value as P_4.

   Set aside
7. Homogenize the mixture in the tube using a vortex for 10 s. Then, let the sample rest for 5 min, repeating this process 6 times, trying to maintain this process within a total interval of 30 min.
8. After resting, transport the tubes containing the starch to the centrifuge (*see* **Note 1**).
9. Centrifuge the tubes at 2200 rpm for 25 min.
10. Immediately after centrifugation, carefully invert the tube over the previously weighed beaker (P_4) and allow the supernatant oil to drain for 2 min.
11. Weigh the beaker containing the drained oil and record the value as P_5.
12. Record all values on the data recording spreadsheet (see Tables 1 and 2 in the **Notes**).
13. Calculate the oil retention capacity as described in the equation in Sect. 3 (*see* **Note 2**).

Table 1
Spreadsheet recording data collected during the analysis

Determination of oil holding capacity (*g* oil / *g* sample)												
Sample												
Analyst												
Date												
Sample identification (code and reference)	Number	Replica	Weight of tube + cap (g) (P_1)	Initial sample weight (g) (P_2)	Weighing tube + sample + added oil (g) (P_3)	Initial weight of tared beaker (g) P_4	Final weight of beaker + drained oil (g) P_5	Oil absorbed by the sample (g)	ORC (g oil /g sample)	Average	D.P.	C.V.
	1	CS1										
CS*	2	CS2										
	3	CS3										

*Example of data recording for a starch sample. *CS* cassava starch, *SD* standard deviation, *CV* coefficient of variation

Table 2
Example of data collected in the analysis

Determination of oil holding capacity (*g* oil / *g* sample)												
Sample		Cassava starch										
Analyst		João										
Date		20/07/2025										
Sample identification (code and reference)	Number	Replica	Weight of tube + cap (g) (P_1)	Initial sample weight (g) (P_2)	Weighing tube + sample + added oil (g) (P_3)	Initial weight of tared beaker (g) P_4	Final weight of beaker + drained oil (g) P_5	Oil absorbed by the sample (g)	ORC (g oil /g sample)	Average	D.P.	C.V.
	1	CS1	12.1916	13.6939	35.7728	10.5246	19.6114	0.8005	0.5328			
CS*	2	CS2	12.2106	13.7187	34.7445	11.1287	19.2185	0.7254	0.4810	0.5127	0.0278	5.4180
	3	CS3	12.2225	13.7261	35.7622	10.5824	19.6078	0.7882	0.5242			

*Example of data recording for a starch sample. *CS* cassava starch, *SD* standard deviation, *CV* coefficient of variation

3.3 Results

Perform the calculation for the calculation of oil absorbed by the sample according to the following equation:

$$Absorbed\ oil\ (g) = (P_3 - P_2 - P_1) - (P_5 - P_4)$$

Where

$$P_1 = Falcon\ tube\ with\ cap,\ in\ g$$
$$P_2 = Falcon\ tube\ with\ cap + sample,\ in\ g$$
$$P_3 = Falcon\ tube\ with\ cap + sample + oil,\ in\ g$$
$$P_4 = Beaker,\ in\ g$$
$$P_5 = Beaker\ with\ drained\ oil,\ in\ g$$

Perform the calculation for the oil holding capacity according to the following equation:

$$Oil\ holding\ capacity\ (g\ oil\ /\ g\ sample) = \frac{Absorbed\ oil\ (g)}{Sample\ weight\ (g)^*}$$

* in dry basis

After obtaining the yield of the three replicates per sample, calculate the mean, standard deviation (SD), and coefficient of variation (CV) according to equations provided by data tabulation programs, such as Microsoft Excel or Google Sheets.

The standard deviation is considered optimal when its value is as close to zero as possible. A coefficient of variation below 10% indicates a dispersion close to ideal for this type of analysis.

4 Notes

1. Fill all capped tubes with the same sample level. Furthermore, the tubes should be placed two at a time (always in pairs) on opposite sides of each other to prevent the centrifuge from becoming unbalanced and potentially stalling.
2. Repeat the analysis in at least triplicate for greater data reliability. Mark the Falcon tube and cap as the same sample set and avoid switching caps, as they may have different masses.

Acknowledgments

The authors would like to thank the National Council for Scientific and Technological Development (CNPq), Brazil, for the scholarships awarded to MTP Clerici (Processes 312660/2023-5). Additionally, the authors express gratitude to the Coordination for the Improvement of Higher Education Personnel (CAPES, financial code 001), Brazil, for the scholarships awarded to NA Brandão (Process 88887.605710/2021-00), HJMC (Process 88887.141456/2025-00) and for supporting the Graduate Program in Food Technology at the Universidade Estadual de Campinas (UNICAMP), São Paulo, Brazil, and the São Paulo Research Foundation (FAPESP), Brazil (Process Number 2023/12726-8) for the postdoctoral fellowship to BL Tagliapietra.

References

1. Lin MJ, Humbert E, Sosulski FW (1974) Certain functional properties of sunflower meal products. J Food Sci 39(2):368–370
2. Carvalho HJM, Barcia MT, Schmiele M (2024) Non-conventional starches: properties and potential applications in food and non-food products. Macromolecules 4(4):886–909
3. Clerici MTPS, Schmiele M (2019) Starches for food application: chemical, technological and health properties, 1st edn. Academic Press, p 433
4. Wurzburg OB (2006) Modified starches. In: Stephen AM, Phillips GO, Williams PA (eds), Food polysaccharides and their applications, 2nd edn. CRC Press, pp 87–112

Check for updates

Chapter 22

Emulsifying Capacity and Stability

Bruna Lago Tagliapietra, Natali Alcantara Brandão, Hugo José Martins Carvalho and Maria Teresa Pedrosa Silva Clerici

Abstract

Emulsifying capacity is a key functional property of starches, enabling their use as natural alternatives to synthetic additives in food formulations. Evaluating this property helps determine starch performance in oil/water systems, including sauces, creams, and emulsified meat products. Both native and modified starches are applied as emulsifiers and as fat and protein substitutes in foods developed for special diets, such as gluten-free, phenylalanine-free, and animal protein-free. These applications support the production of stable and palatable products, including cakes, vegetable creams, sauces, ice cream, and vegan desserts.

Keywords Emulsion, Starch, Foods

1 Introduction

Emulsifying capacity is an important functional property of food ingredients, especially those with the potential to replace synthetic additives in industrial formulations. In the case of starches, assessing this capacity allows us to understand their performance in oil/water dispersion systems, such as sauces, creams, and emulsified meat products.

Native and modified starches have been used as emulsifiers and fat and protein substitutes in food products for special diets, such as gluten-free, phenylalanine-free, animal protein-free, and other diets. Producing foods for these diets requires formulating products that are stable for storage and pleasant to eat, such as cakes, vegetable creams, sauces, ice cream, vegan desserts, and others.

2 Material

2.1 Equipment

1. Analytical balance with precision up to 0.0001 g, within calibration period and with available standard operating procedure.
2. Mixer (speed 8 = approximately 10,000 rpm).

Maria Teresa Pedrosa Silva Clerici and Pedro H. Campelo (eds.), *Basic Protocols in Starch Characterization*, Methods and Protocols in Food Science, https://doi.org/10.1007/978-1-0716-5116-2_22,

3. Vortex.
4. Water bath (30–60°C).
5. Rotating centrifuge for 50 mL Falcon-type tubes with a cooling system.
6. Stopwatch.
7. Millimeter ruler or caliper.

2.2 Material

1. Disposable gloves.
2. Metal spatula.
3. 50 mL Falcon tubes with caps, previously weighed and labeled (tube and cap).
4. Graduated pipettes.
5. Paper towels or tissue for cleaning the outside of the tubes, if necessary.
6. Racks for 50 mL Falcon tubes.

2.3 Reagents

1. Freshly distilled deionized water.
2. Canola oil.

3 Method

Determination of emulsifying capacity and emulsifying stability [1, 2].

3.1 Sample

Dry starch with previously determined moisture content as describe in Chapter 1 of this book.

3.2 Procedure

This protocol aims to determine the emulsifying capacity under low temperature conditions (cold emulsification), using a method adapted from Yasumatsu et al. (1972), as described by Santana et al. (2017).

3.2.1 Preanalysis Steps

1. Preselect the number of tubes according to the number of samples and perform the analyses in triplicate. Check that they are prelabeled and resistant to washing and heating. If not, use a high-temperature pen to identify them.
2. Check the standard operating procedure of the analytical balance and whether it is calibrated. Record the calibration date. Strictly follow the operating procedure.
3. To collect data, prepare a results table in the laboratory notebook, as shown in Table 1.

3.2.2 Emulsifying Capacity

1. Sample weighing: Weigh 1.5000 g of the dry sample and transfer it to a 50 mL Falcon tube.
2. Addition of liquids: Add 15 mL of distilled water and 15 mL of canola oil to the tube.

3. Initial homogenization: Vortex the mixture for 10 s.
   * Ensure that no sample adheres to the bottom of the tube.
4. Intense homogenization: Agitate the mixture using the mixer (speed 8 ≈ 10,000 rpm), making circular movements for 1 min.
5. Initial height (IA) measurement: Before centrifugation, measure the total height of the mixture in the tube using a caliper or ruler and record it as the initial height (IA) in centimeters.
6. Centrifugation: Centrifuge the tubes for 5 min at 3400 rpm (*see* **Note 1**).*
7. Emulsion reading (EA): After centrifugation, identify and measure the height of the formed emulsion layer (EA), located between the lower oil layer and the upper precipitate layer. Record the value in centimeters.
   * The emulsified layer will be located in the middle.
8. Record all values on the data recording spreadsheet (Table 1 in **Notes**).

3.3 Results

Calculate the emulsifying capacity (EC) according to the following equation (*see* **Note 2**):

$$\mathrm{EC}\ (\%) = \frac{\mathrm{EA}}{\mathrm{IA}} \times 100$$

EA = Height of the emulsion layer
IA = Initial height

After obtaining the yield of the three replicates per sample, calculate the mean, standard deviation (SD), and coefficient of variation (CV) according to equations provided by data tabulation programs, such as Microsoft Excel or Google Sheets.

The standard deviation is considered optimal when its value is as close to zero as possible. A coefficient of variation below 10% indicates a dispersion close to ideal for this type of analysis.

3.3.1 Emulsifying Stability

After the initial determination of the emulsifying capacity, the tubes containing the formed emulsion must be subjected to a controlled thermal process to evaluate the stability of the emulsion against heating and cooling.

1. Heating: After step 7 of the emulsifying capacity analysis described above, the Falcon tubes containing the previously formed emulsion should be placed in a water bath at 80°C for a continuous period of 30 min.
2. Cooling: After 30 min, the tubes should be immediately transferred to a container containing water and ice, remaining immersed for 30 min.

3. Centrifugation: Remove the tubes from the water bath and centrifuge for 5 min at 3400 rpm.
   * Remove the tubes from the centrifuge carefully to avoid agitation.
4. Stability reading: Read the height of the emulsion layer remaining in the tube (remaining after heating, cooling, and centrifuging again) using a millimeter ruler or caliper (*see* **Note 2**).
5. Record the remaining emulsion height (REH) values in the data recording spreadsheet (Tables 1 and 2 in **Notes**), which will be used to calculate the emulsion stability, as shown in Fig. 1.

3.3.2 Results

Calculate the emulsifying stability (ES) according to the following equation:

$$\mathrm{ES}\ (\%) = \frac{\mathrm{REH}}{\mathrm{EA}} \times 100$$

EA = Remaining Emulsion Height
EA = Height of the emulsion layer

After obtaining the yield of the three replicates per sample, calculate the mean, standard deviation (SD), and coefficient of variation (CV) according to equations provided by data tabulation programs, as

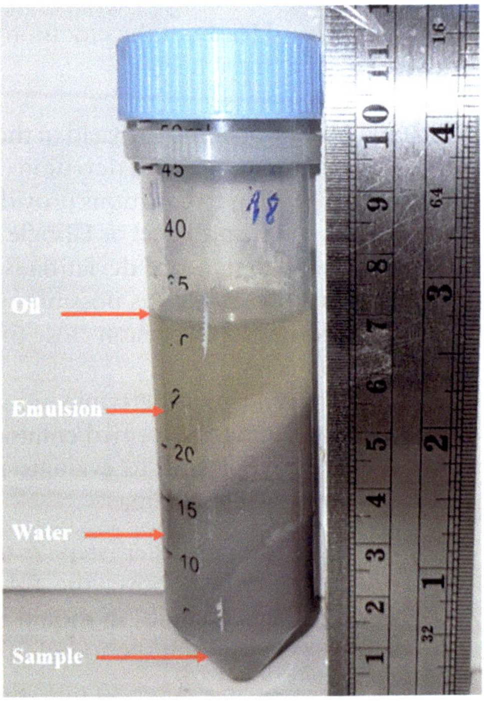

Fig. 1 Emulsion stability and layer formed after procedure

shown in Tables 3 and 4, such as Microsoft Excel or Google Sheets (*see* **Note 3 AND Note 4**).

The standard deviation is considered optimal when its value is as close to zero as possible. A coefficient of variation below 10% indicates a dispersion close to ideal for this type of analysis.

4 Notes

1. Fill all capped tubes with the same sample level. Furthermore, the tubes should be placed two at a time (always in pairs) on opposite sides of each other to prevent the centrifuge from becoming unbalanced and potentially stalling.
2. This analysis measures the cold emulsifying capacity, which will be used in subsequent emulsifying stability (ES) calculations. Repeat the analysis in triplicate for greater data reliability.
3. This methodology was found to be best suited for samples containing a high amount of native starch. Heating the sample before centrifugation (and separating the layers) causes the starch granules to swell and gelatinize, forming a large layer that makes it impossible to separate the gel from the emulsion. In some samples, four layers can be seen after centrifugation.
4. Repeat the analysis in triplicate for greater data reliability.

Table 1
Spreadsheet recording data collected during the analysis

Determination of emulsifying capacity (%)								
Sample								
Analyst								
Date								
Sample identification (code and reference)	Number	Replica	Initial layer height (cm) (A1)	Emulsion layer height (cm) (A2)	CE (%)	Average (%)	DP	CV
	1	CS1						
CS*	2	CS2						
	3	CS3						

*Example of data recording for a starch sample. *CS*, cassava starch; *SD*, standard deviation; *CV*, coefficient of variation

Table 2
Example of data collected in the analysis

Determination of emulsifying capacity (%)								
Sample	Cassava Starch (CS)							
Analyst	Maria							
Date	20/07/2025							
Sample identification (code and reference)	Number	Replica	Initial layer height (cm) (A1)	Emulsion layer height (cm) (A2)	CE (%)	Average (%)	DP	CV
	1	CS1	6.23	1.21	19.4222			
CS*	2	CS2	6.20	1.17	18.8710	18.74	0.76	4.06
	3	CS3	6.25	1.12	17.9200			

*Example of data recording for a starch sample. *CS*, cassava starch; *SD*, standard deviation; *CV*, coefficient of variation

Table 3
Spreadsheet recording data collected during the analysis

Determination of emulsifying stability (%)								
Sample								
Analyst								
Date								
Sample identification (code and reference)	Number	Replica	Emulsion layer height (cm) (A1)	Emulsion layer height after heating (cm) (A2)	ES (%)	Average (%)	DP	CV
	1	CS1						
CS*	2	CS2						
	3	CS3						

*Example of data recording for a starch sample. *CS*, cassava starch; *SD*, standard deviation; *CV*, coefficient of variation

Table 4
Example of data collected in the analysis

Determination of emulsifying stability (%)								
Sample	Cassava starch (CS)							
Analyst	Maria							
Date	20/07/2025							
Sample identification (code and reference)	Number	Replica	Emulsion layer height (cm) (A1)	Emulsion layer height after heating (cm) (A2)	ES (%)	Average (%)	DP	CV
	1	CS1	5.23	0.58	11.0899			
CS*	2	CS2	5.30	0.60	11.3208	11.2020	0.1156	1.0318
	3	CS3	5.27	0.59	11.1954			

*Example of data recording for a starch sample. *CS*, cassava starch; *SD*, standard deviation; *CV*, coefficient of variation

Acknowledgments

The authors would like to thank the National Council for Scientific and Technological Development (CNPq), Brazil, for the scholarships awarded to MTP Clerici (Processes 312660/2023-5). Additionally, the authors express gratitude to the Coordination for the Improvement of Higher Education Personnel (CAPES, financial code 001), Brazil, for the scholarships awarded to HJMC (Process 88887.141456/2025-00) and for supporting the Graduate Program in Food Technology at the Universidade Estadual de Campinas (UNICAMP), São Paulo, Brazil.

References

1. Stone AK, Nosworthy MG, Chiremba C, House JD, Nickerson MT (2019) A comparative study of the functionality and protein quality of a variety of legume and cereal flours. Cereal Chem 96(6):1159–1169
2. Santana GS, de Oliveira Filho JG, Egea MB (2017) Características tecnológicas de farinhas vegetais comerciais. Rev Agric Neotrop 4(2): 88–95

Chapter 23

The Development of Starch Nanofibers by Electrospinning

Laura Martins Fonseca, Elder Pacheco da Cruz, Estefania Julia Dierings de Souza, Alvaro Renato Guerra Dias and Elessandra da Rosa Zavareze

Abstract

The fabrication of starch electrospun fibers is a key area of research for scientists actively developing novel technologies and products utilizing starch. This chapter outlines a general protocol for producing nanofibers via the electrospinning technique. Please note that specific parameters (e.g., polymer concentration, voltage, and flow rate) will vary significantly depending on the source of starch used and the desired properties of the fibers. These parameters are described for each starch source with the corresponding reference that effectively produced the starch nanofibers.

Keywords Electrospinning, Encapsulation, Starch nanofibers

1 Introduction

The electrospinning technique has been established as an efficient strategy for producing polymeric nanofibers with high surface area and highly porous structure, features essential for applications such as the encapsulation and controlled release of bioactive compounds. Starch exhibits promising potential for the production of these nanostructures due to its biodegradability, renewability, and physicochemical properties, which can be tailored to optimize the electrospinning process [1].

The formation of nanofibers via electrospinning depends on the rheological properties of the polymer solution, such as viscosity, surface tension, and shear behavior. Starch, especially after gelatinization, can reach the high viscosities necessary to ensure the molecular entanglement fundamental for continuous fiber formation. Additionally, the shear profile of starch, which tends to be relatively low, favors jet stability during the process, reducing premature breakage and enabling the production of fibers with uniform morphologies.

Maria Teresa Pedrosa Silva Clerici and Pedro H. Campelo (eds.), *Basic Protocols in Starch Characterization*, Methods and Protocols in Food Science, https://doi.org/10.1007/978-1-0716-5116-2_23,

A critical factor in starch electrospinnability is the relative composition of its main components: amylose and amylopectin. Amylose, with its linear structure, facilitates chain entanglement, promoting the formation of fine and continuous nanofibers. On the other hand, amylopectin, which is highly branched, hinders this entanglement, making the production of electrospun nanofibers from starches with low amylose content more challenging. For this reason, starches with higher amylose content are more suitable for electrospinning.

In addition to the intrinsic polymer properties, operational parameters such as solution concentration, applied voltage, needle-to-collector distance, and flow rate play a fundamental role in the quality and characteristics of the produced fibers. Environmental conditions, such as temperature and relative humidity, also influence the jet behavior and, consequently, the morphology of the electrospun nanofibers. This chapter presents a detailed protocol for producing starch nanofibers via electrospinning, covering the necessary equipment, process steps, and optimal operational conditions.

2 Material

2.1 Equipment

The electrospinning apparatus set up (Fig. 1) required for the production of starch nanofibers is relatively simple and comprises the following key components:

- Infusion pump (A): An infusion pump is used to control the flow rate of the polymeric solution.
- Plastic syringe (B): A plastic syringe is coupled to the infusion pump and serves as the reservoir for the polymeric solution. It can be of different capacities, varying from 1 to 20 mL.

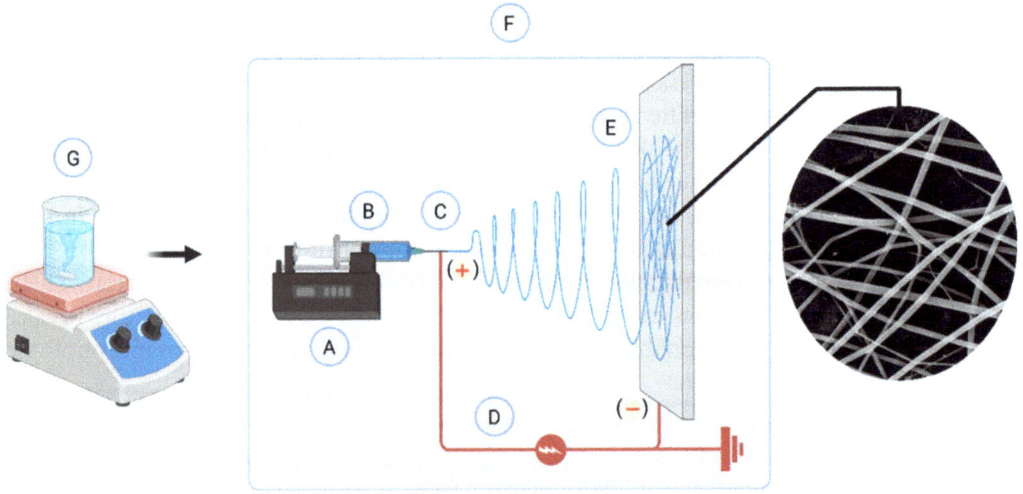

Fig. 1 Electrospinning system for the production of starch nanofibers. Source: Authors' collection, 2025

- Needle (spinneret) (C): A needle, functioning as the spinneret, is attached to the syringe. This is the point from which the polymer jet is ejected. It must be compatible with the syringe.
- High-voltage power supply (D): A high-voltage power supply must be capable of delivering DC voltage in the range of 5–30 kV. It is essential for generating the electric field required for electrospinning. It is connected to both the needle tip and the collector plate.
- Grounded collector plate (E): A grounded conductive plate (e.g., aluminum foil, a rotating drum, or a flat metal sheet) is positioned opposite the needle. This is where the electrospun nanofibers are collected after the solvent evaporates.
- Enclosure/Chamber (F) and dehumidifier: For environmental control, to maintain controlled temperature and humidity (at a maximum of 30 ± 5%).
- For polymeric solution (G) preparation, the polymer, the solvent, a recipient, and a stirrer are needed.

3 Method

3.1 Procedure

The procedure is divided into two main stages: (1) polymer solution preparation and (2) the electrospinning process. Both the solution and process parameters vary for each starch source and are described in Table 1.

3.1.1 Polymer Solution Preparation

First, the optimal starch concentration is determined based on the starch source in the literature or preliminary trials (typically expressed as %w/v or %w/w). This is critical for achieving stable jet formation and the desired fiber morphology. The polymer is then accurately weighed using an analytical balance. In this protocol, we are considering only references that used formic acid as a solvent. Next, accurately measure the required volume of solvent and place the beaker or flask on a magnetic stirrer with a stir bar, or use mechanical stirring. Stir continuously at a moderate speed to ensure complete dissolution. The starch chemical gelatinization is ensured by stirring with formic acid at room temperature for the period determined for each type of starch. The suggested characterization for polymeric solutions includes assessing their viscosity or rheology, surface tension, and electrical conductivity, as these significantly impact the electrospinning process.

3.1.2 Electrospinning Process

First, assemble your electrospinning setup. Carefully load the prepared polymer solution into the syringe, making sure there are no trapped air bubbles. Securely attach the needle to the syringe and

Table 1
Starch and formic acid concentration, starch amylose content, and electrospinning parameters for electrospun fibers production

				Electrospinning parameters			
Starch source	Starch amylose content (%)	Starch concentration (%, w/v)	Formic acid concentration (%, w/v)	Voltage (kV)	Distance needle tip-collector (cm)	Flow rate (mL/h)	Reference
Hylon VII maize starch	70	17	60, 70, 80, 90, and 98	17	11	0.3	[2]
Hylon VII maize	70	25	99	23	20	0.2	[3]
Native Hylon VII (Ingredion)	72	15	75	15	15	0.6	[4]
Native Hylon V (Ingredion)	55	15	75	18	15	0.6	
Anionic Hylon V	–	15	75	18	15	0.6	
Native regular	29.7	12	75	18	15	0.8	
Anionic regular	–	12	75	18	15	0.8	
Soluble potato	32.5	40	75	15	15	0.8	[5]
Yam	33.7	15	75	13.0, 15.8, 20.0, 24.2, 27.0	16.0, 17.8, 20.5, 23.2, 25.0	0.2, 0.36, 0.6, 0.84, 1.0	[6]

Soluble potato	32.5	50	75	25	20	0.6	[7]
Waxy maize (Starpro)	0	17	99	20	15	0.2	[8]
Normal maize (Starpro)	18.8	17	99	20	15	0.2	
High-amylose (Penford)	54.3	17	99	18	15	0.2	
High-amylose (Ingredion)	83.0	17	99	15	15	0.2	
High-amylose (Penford)	69.0	17	99	18	15	0.2	
High-amylose (Roquette)	56.2	17	99	18	15	0.2	
High-amylose (Starpro)	75.2	17	99	16	15	0.2	
Potato (Starpro)	30.5	17	99	16	15	0.2	
Tapioca (Dongmei)	20.7	17	99	17	15	0.2	
Wheat (Dongmei)	24.0	17	99	16	15	0.2	
Green bean (Hada)	34.6	17	99	18	15	0.2	
Cassava (pretreated using reactive extrusion)	49.0	30, 32, 34, and 36	38	19	15	0.2	[9]
Pea (pretreated using reactive extrusion)	43.3	30, 32, 34, and 36	38	19	15	0.2	
Yellow sweet potato	29.7	20	75	18	20	0.75	[10]
White sweet potato	29.8	20	75	18	20	0.75	
Cassava	27.3	20	75	23	20	0.6	[11]

(Continued)

Table 1
continued

Starch source	Starch amylose content (%)	Starch concentration (%, w/v)	Formic acid concentration (%, w/v)	Electrospinning parameters			Reference
				Voltage (kV)	Distance needle tip- collector (cm)	Flow rate (mL/h)	
Taro	12.6	9, 11, 13, 15, and 17	75, 80, 85, 90, 95, and 100	17	15	0.8	[12]
Pea	60.7	10, 20, and 30	75	23	20	0.8	[13]
Bean	49.4	20	75	23	20	1.0	[14]
Avocado seeds	29.7	20	75	23	20	0.6	[15]
Cassava	23.7	15 and 20	75	23	20	0.6	[16]
Soluble potato	24.2	15	75	23	19	0.6	[17]
Yam	33.7	15	75	20	20.5	1.0	[18]

place the filled syringe onto the syringe pump. Connect the high-voltage power supply's positive terminal to the needle tip and the negative terminal to the collector. Verify that the collector is grounded correctly and adjust the distance between the needle tip and the collector. Close the electrospinning chamber and allow the humidity to stabilize to the desired levels.

Initiate the electrospinning process by setting your desired flow rate and starting the pump. Confirm that a small droplet forms at the needle tip. Gradually increase the high voltage from the power supply and observe the formation of a Taylor cone (a conical shape) at the needle tip. As the voltage increases, the Taylor cone will elongate, and a stable jet of solution should emerge. Throughout the process, continuously monitor for stability, proper Taylor cone formation, and consistent fiber deposition. Use a light source if needed. Continue electrospinning for the desired duration to collect sufficient material.

3.1.3 Postprocessing

Once spinning is complete, postprocessing begins. First, turn off the high-voltage power supply. Then, stop the syringe pump and carefully remove the fibrous mat from the collector. Store the dried fiber mats in a dry, sealed container to prevent moisture absorption or degradation. The suggested basic characterization for the starch nanofibers is morphology and size distribution.

Acknowledgments

The authors thank CAPES (code 001).

References

1. Fonseca LM, Cruz EP, Crizel RL, Jansen-Alves C, Dias ARG, Zavareze ER (2024) New advances of electrospun starch fibers, encapsulation, and food applications: a review. Trends Food Sci Technol 147:104467
2. Lancuški A, Vasilyev G, Putaux JL, Zussman E (2015) Rheological properties and electrospinnability of high-amylose starch in formic acid. Biomacromolecules 16(8):2529–2536
3. Wu D, Samanta A, Srivastava RK, Hakkarainen M (2017) Starch-derived nanographene oxide paves the way for electrospinnable and bioactive starch scaffolds for bone tissue engineering. Biomacromolecules 18(5):1582–1591
4. Fonseca LM, Oliveira JP, Oliveira PD, Zavareze ER, Dias ARG, Lim LT (2019) Electrospinning of native and anionic corn starch fibers with different amylose contents. Food Res Int 116:1318–1326
5. Fonseca LM, Silva FT, Antunes MD, Halal SLM, Lim LT, Dias ARG (2019) Aging time of soluble potato starch solutions for ultrafine fibers formation by electrospinning. Starch/Staerke 71(1–2):1800089
6. Souza EJD, Santos FN, Pires JB, Kringel DH, Silva WMF, Meinhart AD, Dias ARG, Zavareze ER (2021) Production and optimization of ultrafine fiber from yam starch by electrospinning method using multivariate analysis. Starch/Staerke 73(3–4):2000174
7. Cruz EP, Fonseca LM, Radünz M, Silva FT, Gandra EA, Zavareze ER, Borges CD (2021) Pinhão coat extract encapsulated in starch ultrafine fibers: thermal, antioxidant, and antimicrobial properties and in vitro biological digestion. J Food Sci 86(7):2886–2897
8. Cao P, Wu G, Yao Z, Wang Z, Li E, Yu S, Liu Q, Gilbert RG, Li S (2022) Effects of amylose and amylopectin molecular structures on starch electrospinning. Carbohydr Polym 296:119959
9. Tochihuitl-Vázquez D, Ramírez-Bon R, Yáñez-Limón JM, Martínez-Bustos F (2022) Reactive

extrusion as a pretreatment in cassava (*Manihot esculenta* Crantz) and pea (*Pisum sativum* L.) starches to improve spinnability properties for obtaining fibers. Molecules 27(18):5944

10. Cruz EP, Jansen ET, Fonseca LM, Hackbart HCS, Siebeneichler TJ, Pires JB, Gandra EA, Rombaldi CV, Zavareze ER, Dias ARG (2023) Red onion skin extract rich in flavonoids encapsulated in ultrafine fibers of sweet potato starch by electrospinning. Food Chem 406:134954
11. Cruz EP, Pires JB, Santos FN, Fonseca LM, Radünz M, Dal Magro J, Gandra EA, Zavareze ER, Dias ARG (2023) Encapsulation of lemongrass essential oil into cassava starch fibers for application as antifungal agents in bread. Food Hydroc 145:109105
12. Sawitri A, Luthfianti HR, Hapidin DA, Edikresnha D, Palar PS, Surtiyeni N, Wibowo A, Khairurrijal K (2024) Synthesis and characterization of novel electrospun nanofibers based on taro starch: influence of solvent and isolation agent on morphology and diameter. Polym Int 74(3):217–230
13. Cruz EP, Santos FN, Souza JF, Souza EJD, Fonseca LM, Fajardo AR, Zavareze ER, Dias ARG (2024) Impact of concentration and aging time of pea starch-based polymeric solutions on the fabrication of electrospun nanofibers. Polym Adv Technol 35(6):e6479
14. Santos FN, Fonseca LM, Jansen-Alves C, Crizel RL, Pires JB, Kroning IS, de Souza JF, Fajardo AR, Lopes GV, Dias ARG, Zavareze ER (2024) Antimicrobial activity of geranium (*Pelargonium graveolens*) essential oil and its encapsulation in carioca bean starch ultrafine fibers by electrospinning. Int J Biol Macromol 265(1):130953
15. Pires JB, Santos FN, Cruz EP, Fonseca LM, Siebeneichler TJ, Lemos GS, Gandra EA, Zavareze ER, Dias ARG (2024) Starch extraction from avocado by-product and its use for encapsulation of ginger essential oil by electrospinning. Int J Biol Macromol 254(1):127617
16. Santos FN, Cruz EP, Fonseca LM, Pires JB, Diaz PS, Dias ARG, Zavareze ER (2024) Cassava starch esterification with formic acid for fabrication of electrospun fibers. Int J Biol Macromol 266(2):131182
17. Costa IHL, Hackbart HC, Oliveira G, Pires JB, Filho PJS, Weber FH, Borges GSC, Zavareze ER, Dias ARG (2024) Pitaya (*Hylocereus polyrhizus*) extract rich in betanin encapsulated in electrospun sweet potato starch nanofibers. J Sci Food Agric 105(2):878–889
18. Souza EJD, Santos FN, Pires JB, Cruz EP, Fonseca LM, Gandra EA, Dias ARG, Zavareze ER (2025) Potential use of yam starch: thermal protection of tea tree essential oil encapsulated by electrospinning. Starch/Staerke 77(5):e70000

Check for updates

Chapter 24

The Development of Starch Nanocapsules by Electrospraying

Elder Pacheco da Cruz, Laura Martins Fonseca, Estefania Julia Dierings de Souza, Elessandra da Rosa Zavareze and Alvaro Renato Guerra Dias

Abstract

For scientists developing new starch-based technologies, the fabrication of starch electrosprayed capsules is an innovative research area. This chapter outlines a general protocol for their production via electrospraying, noting that specific parameters (e.g., polymer concentration, voltage, and flow rate) vary. We describe these parameters for each starch source, including references for their effective use in producing starch nanocapsules.

Keywords Encapsulation, Starch-based materials, Biopolymers, Food technology

1 Introduction

Starch is a biodegradable and renewable polymer with significant potential for encapsulation of bioactive compounds via electrospraying, owing to its favorable physicochemical properties, structural versatility, and broad availability. Electrospraying has gained attention for encapsulating bioactive compounds, particularly due to its ability to operate under mild conditions that help preserve the integrity of temperature–sensitive molecules. It enables the production of nano and microparticles that protect active compounds from degradation caused by light, oxygen, pH variations, and enzymatic activity. Moreover, electrospraying offers advantages such as enhanced bioavailability, controlled release, and masking of undesirable sensory attributes, making it highly attractive for applications in food and pharmaceutical formulations.

The electrospraying system consists of four main components: a high-voltage power supply, a syringe pump, a metallic needle, and a grounded collector. The technique is based on applying a high electric field to a polymer solution, which is then ejected toward the collector, promoting the formation of spherical nanocapsules with

Maria Teresa Pedrosa Silva Clerici and Pedro H. Campelo (eds.), *Basic Protocols in Starch Characterization*, Methods and Protocols in Food Science, https://doi.org/10.1007/978-1-0716-5116-2_24,

various morphologies. Although the technical terms are "spheres" or "particles," the literature more commonly refers to these materials as "capsules."

Starch has emerged as a versatile material for encapsulation systems, particularly due to its functional and structural characteristics that enable effective interaction with active compounds. Derived from plant sources, starch also exhibits specific factors, such as tunable solubility, enzymatic digestibility, and rheological properties, that support its use in controlled-release systems. However, despite its broad application in modified forms, producing capsules via electrospraying using native starch remains a technical challenge, with only a limited number of studies available in the literature. Variables such as the starch source, its concentration, and the treatment of the polymer solution (e.g., aging time after gelatinization) directly affect the success of the process.

The formation of electrosprayed capsules relies on a complex set of intrinsic and extrinsic variables. Factors such as polymer concentration, solution viscosity, electrical conductivity, and surface tension, as well as operational parameters like applied voltage, needle-to-collector distance, needle diameter, and pump flow rate, must be carefully optimized for each type of polymer. Furthermore, environmental conditions such as temperature and relative humidity also have a direct impact on the behavior of the solution within the electric field and, consequently, on the morphology and final properties of the capsules produced.

This chapter presents the protocol for producing starch nanocapsules via electrospraying, discussing the necessary equipment, procedures, and process conditions while considering the multiple variables involved, such as starch concentration, solvent type and concentration, starch amylose content, and electrospraying parameters for the production of electrospun capsules.

2 Material

2.1 Equipment

The electrospraying apparatus set up (Fig. 1) required for the production of starch nanocapsules is relatively simple and comprises the following key components:

- Infusion pump (A): An infusion pump is used to control the flow rate of the polymeric solution.
- Plastic syringe (B): A plastic syringe is coupled to the infusion pump and serves as the reservoir for the polymeric solution. It can be of different capacities, varying from 1 to 20 mL.
- Needle (spinneret) (C): A needle, functioning as the spinneret, is attached to the syringe. This is the point from which the polymer jet is ejected. It must be compatible with the syringe.

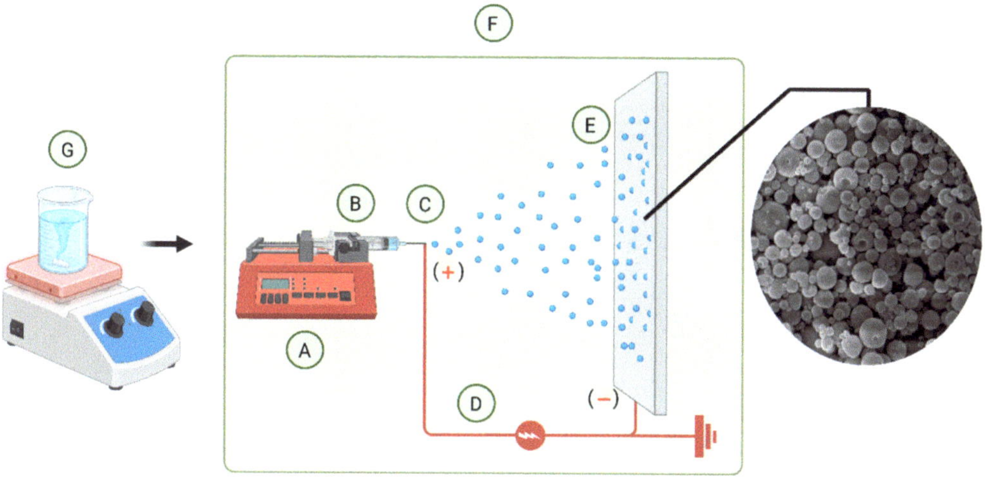

Fig. 1 Electrospray system for the production of starch nanocapsules. Source: Authors' collection, 2025

- High-voltage power supply (D): A high-voltage power supply must be capable of delivering DC voltage in the range of 5–30 kV. It is essential for generating the electric field required for electrospraying. It is connected to both the needle tip and the collector plate.
- Grounded collector plate (E): A grounded conductive plate (e.g., aluminum foil, a rotating drum, or a flat metal sheet) is positioned opposite the needle. This is where the electrosprayed nanocapsules are collected after the solvent evaporates.
- Enclosure/chamber (F) and dehumidifier: For environmental control, to maintain controlled temperature and humidity (at a maximum of 30 ± 5%).
- For the preparation of a polymeric solution (*G*), the following materials are required: the polymer, the solvent, a recipient, and a stirrer.

3 Method

3.1 Procedure

The procedure is divided into two main stages: (1) polymer solution preparation and (2) the electrospraying process. Both the solution and process parameters vary for each starch source and are described in Table 1.

3.1.1 Polymer Solution Preparation

First, the optimal starch concentration is determined based on the starch source in the literature or preliminary trials (typically expressed as %w/v or %w/w). The solution for capsule production is not too

Table 1
Starch concentration, solvent type and concentration, starch amylose content, and electrospraying parameters for electrosprayed capsules production

Starch source	Starch amylose content (%)	Ideal starch concentration (%, w/v)	Solvent and concentration (%, v/v)	Stirring time	Electrospraying parameters			Reference
					Voltage (kV)	Distance needle tip-collector (cm)	Flow rate (mL/h)	
OSA-starch CAPSUL®	–	20	Ethanol solution—20, 30, or 40	–	15	20	0.6	[1]
Potato	–	10	Formic acid—75	24 h	23	15	0.6	[2]
OSA-starch INGREDION®	–	40	Ethanol solution—20	30 min	23	15	0.6	[3]
Cassava	23.4	7	Formic acid—75	24 h	23	15	0.6	
Corn	25.7	7	Formic acid—75	24 h	23	15	0.6	[4]
Wheat	30.5	7	Formic acid—75	24 h	23	15	0.6	
Sweet potato	24.9	7	Formic acid—75	24 h	23	15	0.6	
OSA-starch INGREDION®	–	40 and 50	Ethanol solution—20 and 30	30 min	23	15	0.6	[5]

OSA-starch octenyl succinic anhydride starch

viscous. The polymer is then accurately weighed using an analytical balance. In this protocol, we consider both native starches, which use formic acid as a solvent, and modified starchesmodified, which use ethanol as a solvent. Next, accurately measure the required volume of solvent and place the beaker or flask on a magnetic stirrer with a stir bar, or use mechanical stirring. Stir continuously at a moderate speed to ensure complete dissolution. The starch chemical gelatinization is ensured by stirring with formic acid at room temperature for the period determined for each type of starch. If modified starches are used, ethanol is responsible for dissolving the starch and forming the polymer solution. The suggested characterization for polymeric solutions includes assessing their viscosity or rheology, surface tension, and electrical conductivity, as these significantly impact the electrospraying process.

3.1.2 Electrospraying Process

First, assemble your electrospraying setup. Carefully load the prepared polymer solution into the syringe, making sure there are no trapped air bubbles. Securely attach the needle to the syringe and place the filled syringe onto the syringe pump. Connect the high-voltage power supply's positive terminal to the needle tip and the negative terminal to the collector. Verify that the collector is grounded correctly and adjust the distance between the needle tip and the collector. Close the chamber and allow the humidity to stabilize to the desired levels.

Initiate the electrospraying process by setting your desired flow rate and starting the pump. Confirm that a small droplet forms at the needle tip. Gradually increase the high voltage from the power supply and observe the pulverization of the solution from the needle tip. Throughout the process, continuously monitor for stability and consistent capsule deposition. Use a light source if needed. Continue electrospraying for the desired duration to collect a sufficient amount of material.

3.1.3 Postprocessing

Once spraying is complete, postprocessing begins. First, turn off the high-voltage power supply. Then, stop the syringe pump and carefully remove the mat of capsules from the collector. Store the dried material in a dry, sealed container to prevent moisture absorption or degradation. The suggested basic characterization for the starch nanocapsules is morphology and size distribution.

Acknowledgments

The authors thank CAPES (code 001).

References

1. Biduski B, Kringel DH, Colussi R, Hackbart HCS, Lim LT, Dias ARG, Zavareze ER (2019) Electrosprayed octenyl succinic anhydride starch capsules for rosemary essential oil encapsulation. Int J Biol Macromol 132:300–307
2. Pires JB, Fonseca LM, Siebeneichler TJ, Crizel RL, Santos FN, Hackbart HCS, Kringel DH, Meinhart AD, Zavareze ER, Dias ARG (2022) Curcumin encapsulation in capsules and fibers of potato starch by electrospraying and electrospinning: thermal resistance and antioxidant activity. Food Res Int 162(Part B):112111. 162
3. Souza EJD, Kringel DH, Siebeneichler TJ, Rombaldi CV, Gandra EA, Dias ARG, Zavareze ER (2024) Antifungal potential of clove and white thyme essential oils encapsulated in electrosprayed starch-OSA against *Penicillium digitatum*. Eur Polym J 215:113192
4. Pires JB, Santos FN, Cruz EP, Fonseca LM, Pacheco CO, Rosa BN, Santana LR, Pereira CMP, Carreno NLV, Diaz PS, Zavareze ER, Dias ARG (2024) Cassava, corn, wheat, and sweet potato native starches: a promising biopolymer in the production of capsules by electrospraying. Int J Biol Macromol 281(3):136436. 281
5. Souza EJD, Santos FN, Kringel DH, Souza JF, Bueno DT, Leitzke AF, Fajardo AR, Gandra EA, Pereira CMP, Dias ARG, Zavareze ER (2024) Multivariate optimization for the production of OSA-starch microcapsules containing clove and white thyme essential oils: antifungal potential against *Penicillium digitatum*, 215. Polym Adv Technol 35(12):113192

INDEX

Maria Teresa Pedrosa Silva Clerici and Pedro H. Campelo (eds.), *Basic Protocols in Starch Characterization*, Methods and Protocols in Food Science, https://doi.org/10.1007/978-1-0716-5116-2,

MIX
Papier aus verantwortungsvollen Quellen
Paper from responsible sources
FSC® C105338

If you have any concerns about our products,
you can contact us on
ProductSafety@springernature.com

In case Publisher is established outside the EU,
the EU authorized representative is:
Springer Nature Customer Service Center GmbH
Europaplatz 3, 69115 Heidelberg, Germany

Printed by Libri Plureos GmbH
in Hamburg, Germany